Spotlight SCIENCE 8

Keith JOHNSON ★ **Sue ADAMSON** ★ **Gareth WILLIAMS** ★ **Lawrie RYAN**

With the active support of: Bob Wakefield, Roger Frost, Helen Davis, Phil Bunyan, Michael Cotter, Cathryn Mellor, Peter Borrows, John Bailey, Janet Hawkins, Sarah Ryan, Judy Ryan, Ann Johnson, Graham Adamson, Diana Williams.

FRAMEWORK EDITION

Nelson Thornes

First published in 2003 by:
Nelson Thornes Ltd
Delta Place
27 Bath Road
CHELTENHAM
GL53 7TH
United Kingdom

08 / 10 9 8 7 6 5

A catalogue record for this book is available from the British Library

ISBN 978 0 7487 7473 9

Illustrations by Jane Cope, Angela Lumley, Mike Gordon and Peters & Zabransky
Page make-up by Tech-Set

Printed and bound in China by Midas

Acknowledgements

The authors and publishers are grateful to the following for permission to reproduce photographs:

A1 pix Superbild: Ro-Ma Stock 27, H Smidbauer 43, Incolor 162TR; Ace Picture Library: Photolibrary International 101M; Action Plus:
Glyn Kirk 35; Adams Picture Library: 144TL, TR; AES Educational: 31d; Al Hamdan: 31BL; Allsport: David Cannon 135; A Muttit:
55B; Ann Ronan Picture Library: 61B, 80B, 81, 116TL, 147T; Aquarius Picture Library: 20; Axon Images: 82ML, 104T; Bart's Medical
Library: 21R; Biophoto Associates: 33B, 34B, 50B; BOC Gases: 76B; Bodleian Library: 56R, L; BP Amoco: 82T; Bridgeman Art Library:
Edelfelt AGA 29; British Airways: 65TL; Bruce Coleman: 54, Felix Labhardt 48B, Jane Burton 48T; Bruno Gardent: 13; Camera Press
London: 119e, Karsh of Ottowa 147B; Corbis: Lester Lefkowitz 31f, Paul A Souders 49B; Corel (NT): 48MT, 50T, 51, 58, 69B, 75B,
78B, 79T, MT, MB, 82MR, 84, 88BL, 89TL, TR, B, 90BL, 95, 103B, 105B, 112C, 128L; Dairy Council: 31b; Digital Vision (NT): 6B, 31a,
45B, 63, 100, 101B; Dr Eckart Pott: 45T; Ecoscene: Anthony Cooper 104BL; Eye Ubiquitous: James Davis Worldwide 83; Fisons:
158BL; Frank Greenaway: 151R, 162ML; Frank Lane Picture Agency: W Howes 45M, M Newman 86B, M Walker 90BR, Celtic Picture
Agency 97B, R Wilmshurst 110BR, 151L, 162BL, Tom & Pam Gardener 131; Gary M Prior: 16T, 19; Geological Museum London: 66T;
Geophotos: 102T, M, 103T, 106MR; Geoscience Features Picture Library: 66MC, 72R, 73TC, B, 85TR, ML, MR, 96TL, TR, BL, BR,
97TL, TR, 101TL, TC, TR, 102BL, BR, 98B, 106ML, MC, 105MT; Getty Images Photodisc: Squared Studios 119b; Getty Images Stone:
David Joel/McNeal Hospital 162BR; Getty Images Taxi: L Lefkowitz 129; Geheime Denmark: 116B; GT Insulations: 110BL; Highgrove
Foods: 74M; Holt Studios International: 30T, 158BR; Hulton Getty: 41, 117, 163; ICI: 47B, 78T, 105MBC; Image Library (NT): 17, 60TR;
Image State: 71; James Longley: 104BR; Jane Burton: 85TL; John Cancalosi: 93B; John Walmsley: 4; J Moss: 170M; Jules Cowan:
44BL; J V DeFord Jr: 112R; K Johnson: 134; Kanehara & Co. Ltd: 144B (Ishihara's tests for colour blindness cannot be conducted
with this material); Last Resort Picture Library: 48MB, 79M, 87, 92; Martyn Chillmaid: 6R, L, 28B, 33T, 47T, 49T, M (PH meter courtesy
of Hanna Instruments UK Ltd), 60TL, B, 61TL, TCL, TCR, TR, M, 64, 65TC, 66ML, MR, B, 67BL, BC, BR, 68TL, TR, B, 69TL, TR, 72L,
73TR, 74BC, 75TR, TL, B, 76T, 77, 79BL, 80T, 82MCL, 110T, 113BR, 114TL, TR, TC, BL, BC, BR, 118, 119a, c, 124a, b, 146B, 156,
160; Mary Evans Picture Library: 67T, 116TR; Michael Scott: 7; Mike Read: 52; Moorfields Eye Hospital: 124c; NASA: 162TL; National
Blood Service: 26B; Natural History Museum London: 99B; Natural Visions: 44BR, 55TL, TR; Neil McAllister: 85B; Nicholas De Vore:
74BL; Nokia: 119d; Oxford Scientific Films: Tim Shepherd 44T, P Gathercole 55MR, Colin Milkins 55BL, J A L Cooke 55BR, Breck
P Kent 105T; Photodisc (NT): 90M, 140, 150; Photographers Library: 3TL; Quadrant Picture Library: 158T; Robert Harding Picture
Library: 36B, 105MBL, 107Yoav Levy/Phototake 40B, T Waltham 98TL, M Jenner 98TR; Rod Williams: 153; Roger Labrosse: 26T; Royal
Society of Chemistry: 60TC; Science & Society Picture Library: 99T, 116M; Science Photolibrary: 23, 34T, 37, 40M, Sheila Terry 6TC,
Larry Mulverhill 13inset, John Radcliffe Hospital 22, Sidney Moulds 28TR, Claude Nuridsany & Marie Perennou 28TL, David Hall 31e,
CNRI 32L, A B Dowsett 32R, Andrew Lambert Photography 70T, 73TL, Klaus Gildbrandsen 74BR, Adam Hart 86TL, Michael Marten
88T, Sinclair Stammers 93T, 98ML, Louise K Broman 98TC, Mehan Kulyk 98MR, G Williams 113BL, Simon Fraser/Department of
Neuroradiology/Newcastle General Hospital, 119f, 126, Alex Bartel 124d, Oscar Burriel/Latin Stock 130, Gordon Garradd 142, Hattie
Young 152T; Science Pictures: 21L; Shout Pictures: 146TL, TR; Stephen Munday: 16B; Stockbyte (NT): 31c, 90T; T Clayton: 79BR;
Topham Picture Point: 110BC; Wadworth & Co. Ltd: 30B; World Gold Council: 65TR; Picture research by johnbailey@axonimages.com

Contents

Getting enough energy?

Learn about:
● energy from foods
● different energy needs

Think of the ways in which your body uses up energy.

▶ Make a list of things you have done today which have used up some of your energy.

Energy intake

Sue is a 12-year-old girl. The average energy needed by a girl her age is 9700 kJ per day.

▶ Look at Sue's meals for the last 24 hours and answer these questions.

a Work out Sue's total energy intake for the day (in kJ).

b Did Sue get enough energy for a girl of her age?

c Which 5 foods gave her most energy?

d What do you think would happen if her energy intake was much lower than 9700 kJ per day? How would this make her feel when she needed to be active?

Sue's food	Energy in kJ
cereal	400
milk	600
choc. bar	1500
cheese sandwich	1800
crisps	600
lemonade	700
pizza	1200
chips	1000
apple pie	1200
custard	600
cup of tea	200

How much energy?

▶ Now work out your energy intake for the food you ate during the last 24 hours. Use this table to find the energy content of each food, in kilojoules (kJ).

Average portion	kJ	Average portion	kJ	Average portion	kJ
red meat	2000	cornflakes	400	apple	200
chicken	900	milk (1 cup)	600	apple pie	1200
beefburger	1700	yoghurt	400	banana	300
lamb curry	1200	cheese	900	orange	200
pizza	1200	peas	300	chapatti	900
sausages	1500	tomatoes	100	bread (1 slice)	400
fish fingers	700	cabbage	80	pat of butter	200
spaghetti	500	carrots	80	jam	400
rice (boiled)	500	lettuce	40	cake (1 slice)	700
potatoes, boiled	400	choc. bar	1500	cola	600
chips	1000	ice cream	500	coffee	100
baked beans	400	crisps	600	sugar (teaspoon)	100
egg, boiled	400	jelly	300	squash	300
fried	500	biscuit	400	thick soup	600

e Did you eat enough to cover your energy needs?
(12-year-old girl = about 9700 kJ per day;
12-year-old boy = about 11 700 kJ per day)

f What advice would you give someone who is overweight about:
i) taking in less energy?
ii) using up more energy?

Different energy needs

The amount of energy that you need depends on:
- how big you are,
- how active you are,
- how fast you are growing.

▶ Look at the pictures and then answer these questions:

g Why do you think that males usually need more energy than females?

h Why do manual workers need more energy than office workers? How much more energy?

i Why does a 13-year-old boy need more energy than a male office worker? How much more does he need?

j Why does pregnancy increase a woman's energy needs?

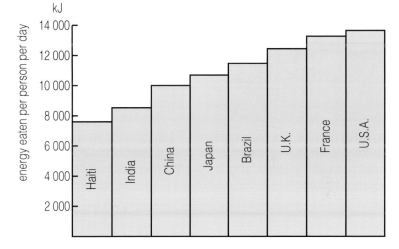

boy 12–15 years
11 700 kJ/day

male manual worker
15 000 kJ/day

female office worker
9800 kJ/day

girl 12–15 years
9700 kJ/day

male office worker
11 000 kJ/day

pregnant woman
10 000 kJ/day

Energy intakes in different countries

▶ Look at the bar-chart:

k Which of these countries eats the most energy foods per person?

l Which country eats the least?

m Why do you think there are such large differences?
How will this affect the people's health?
Discuss these questions within your group.

kJ

energy eaten per person per day

14 000 — 12 000 — 10 000 — 8 000 — 6 000 — 4 000 — 2 000 —

Haiti · India · China · Japan · Brazil · U.K. · France · U.S.A.

---

1 Draw a bar-chart to show the energy needed per minute for these activities:

sleeping	4 kJ per minute
eating	6 kJ per minute
writing	7 kJ per minute
walking	15 kJ per minute
climbing stairs	20 kJ per minute
running	30 kJ per minute

2 Use the data in the table on the opposite page to plan:
a) a meal to give you about 3500 kJ
b) a day's diet for a female office worker who wants to lose weight
c) a day's diet for a male distance-runner in training for a race.

3 Make a survey of your class to find out what they eat for lunch.
Can you see any pattern in your results?

4 Find out about the dangers of slimming too much. What are **anorexia** and **bulimia**?

5 Plan an investigation to find out which foods birds prefer.

6 Use the internet or your school intranet to search for information about "world energy consumption" or "world energy resources". Write a brief report about the most important piece of information that you find. Explain why you think it is important.

Things to do

Food for thought

Do you have any favourite foods?

▶ Make a list of some of the foods that you like to eat.

Which of these foods do you think are good for you?

Draw a ring around those foods that are not good for you.

Healthy eating

You need a healthy diet to:
• grow • repair damaged cells • get energy • keep healthy.
A healthy diet should include some of each of these:

Proteins are for growth. They are used to make new cells and repair damaged tissue.

Carbohydrates, like sugar and starch, are our high-energy foods, but eat too much and they turn to fat.

Fats are used to store energy. They also insulate our bodies so that we do not lose a lot of heat.

Vitamins and minerals are needed in small amounts to keep us healthy e.g. iron for the blood and vitamin D for the bones.

The easiest way to have a healthy diet is to eat a variety of foods each day.

▶ Choose one food from each picture to plan some healthy meals.

▶ Look at the questions opposite:

Choose one of these questions, or one of your own, to answer. You can use reference books, videos, ROMs, information booklets or the internet to conduct your research.
Present your findings in a way that will appeal to others in your class. Point out which parts are facts and which are opinions.

How do diets of different cultures differ?

What foods should you eat to reduce the chance of heart disease?

What are organic foods and how are they different from other foods?

Are breakfast cereals really good for you?

Should children and adults have the same diet?

Food tests

Here are 4 ways of testing for foods.

Do each test carefully and observe the result.

Write down your results in each case.

Testing for starch

Add 2 drops of *iodine* **solution** to some starch solution.

What do you **see?**

eye protection

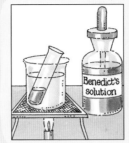

Testing for glucose

Add 10 drops of **Benedict's solution** to one-fifth of a test-tube of glucose solution.

Heat carefully in a water bath.

What do you **see?**

eye protection

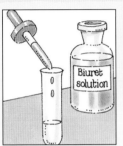

Testing for protein

Add 10 drops of **Biuret solution** (be careful: this is an irritant) to half a test-tube of protein solution.

What do you **see?**

⚠

eye protection
Biuret solution
is an irritant

Testing for fat

Rub some of the food onto a piece of filter paper.

Hold the paper up to the light.

What do you **see?**

Now try testing a few foods. If the food is a solid you will have to grind it up with a little water first.

Record your results in a table like this:

Food	Starch	Glucose	Protein	Fat
Nuts			✓	✓

Produce a Venn diagram to display the data in your table.

Things to do

1 Copy and complete the table:

Food	Use to my body	Food containing a lot of it	Chemical test
protein starch fat glucose			

2 Do some research to find out what these vitamins and minerals are needed for. What happens if you do not get enough of them?
a) vitamin C
b) iron
c) vitamin A
d) calcium
e) vitamin B group
f) iodine.

3 Keep a careful record of all the food that you eat in the next 24 hours.
Use the Recommended Daily Amounts table to see if you had a healthy diet.

4 Fibre or roughage is an essential part of your diet. Do some research to find out
a) what fibre is   b) where it is found
c) why it is so important to us.

8A3

Digestion

Learn about:
● digestion of food
● how enzymes work
● fibre in our diet

Think of the different foods that you eat.
How much of your food is **soluble** (will dissolve in water)**?**
Probably not much.

Before our bodies can use the food that we eat it must be **digested**.
When food is digested it is broken down into very small molecules:

There are special **digestive juices** in our body.
These digest large molecules into small ones.
(You can read more about molecules on page 62.)

▶ Try chewing some bread for a long time. Eventually it tastes sweet
because your saliva has broken down the starch in the bread to sugar.

Starch is a very big molecule. It is made up
of lots of sugar molecules joined together.

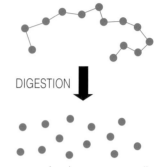

DIGESTION

sugar molecules are very small

How enzymes work

enzymes cut
up starch

starch
molecule

sugar molecules

Saliva contains a chemical called an **enzyme**.
This acts like scissors to cut up the starch molecule into sugar
molecules.

Proteins are large food molecules.
They are made up of small **amino acids** joined together.

Fats are large food molecules.
They are made up of small **fatty acids** joined together.

a What do you think happens to proteins and fats in digestion**?**
You could draw simple diagrams to help your explanation.

Enzymes can be very particular.
For instance, the enzyme in saliva will only digest starch.
It will not digest protein or fat.
Similarly it takes a different enzyme to digest protein and another
one to digest fat.
For this reason we say that enzymes are *specific*.

The small molecules made when you digest food
are transported around your body in your blood.
This is how they can reach every cell in your body.

Changing starch into sugar

You can find out how saliva affects starch by carrying out this experiment.

1 Set the 2 test-tubes up as shown in the diagram.
2 Leave the apparatus for 10 minutes at 40 °C.
3 Test a drop from each test-tube for starch with iodine. What do you see?
4 Add a small amount of Benedict's solution to each test-tube and test for sugar. What do you see?
5 Record your results in a table like this:

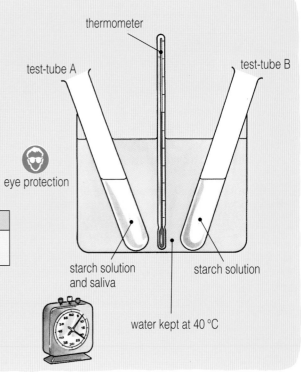

thermometer

test-tube A

test-tube B

eye protection

starch solution and saliva

starch solution

water kept at 40 °C

	Colour with iodine	Colour with Benedict's
test-tube A test-tube B		

b In which test-tube was the starch broken down?
c What do you think the starch was broken down to?
d Why do you think test-tube B contained only starch solution?
e Why were your test-tubes kept at 40 °C?

Why digest?

Food must be broken down and made soluble in your body.
This is so that it can pass through your gut wall into your blood.
Your blood then carries the digested food all around your body.

f Why can't undigested food pass through your gut wall into your blood?
g Which parts of your body need the food?
h In what ways does your body 'use up' food?

Fibre cannot be digested so it isn't broken down.
Fibre adds bulk and a solid shape to your food so that it can be pushed along your gut.
Tough, stringy plants like celery have lots of fibre.

▶ Make a list of foods that are high in fibre.

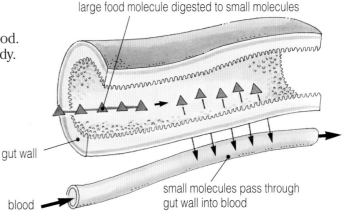

large food molecule digested to small molecules

gut wall

blood

small molecules pass through gut wall into blood

1 Copy and complete:
Digestion is the .... down of food into .... molecules by chemicals called .... Food has to be digested so that it can pass through the .... wall into the .... stream. Starch is digested to ...., protein is digested to .... acids and fats are digested to .... acids.

2 Biological washing powders contain enzymes. Many stains contain fat and protein that can be digested by enzymes. Plan an investigation into the effect of temperature on biological and non-biological washing powders.

3 The graph shows the effect of temperature on the activity of an enzyme.
a) At what temperature is the enzyme most active?
b) Explain what is happening to the action of the enzyme i) between X and Y ii) between Y and Z.

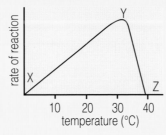

rate of reaction

temperature (°C)

Things to do

Your gut

What happens to your food when you swallow**?**
It enters a tube that starts with your mouth and ends at your anus.
The whole of this food tube is called your **gut**.

Your gut is about 9 metres long.

▶ Work out how many times your height it is.

a How does all this length of gut fit into your body**?**

b Why do you think your gut has to be so long**?**

Look at the diagram of the human gut below.

▶ Follow the path your food goes down.
There are lots of twists and turns.

c Write down the correct order of parts that food passes through.

Down the tube

Mouth
Food chewed and mixed with saliva.
Then you swallow it (gulp!).
(Food is here for 20 seconds)

Gullet
A straight, muscular tube leading to your stomach.
(10 seconds)

Stomach
The acid bath! Digestive juices and acid are added to food
here. Your stomach churns up this mixture.
(2 to 6 hours)

Small intestine
More juices are added from your liver and your pancreas.
These complete digestion. Then the food passes through
into your blood. This is called **absorption**.
(About 5 hours)

Large intestine
Only food that can not be digested (like fibre) reaches here.
A lot of water passes back into your body. This leaves solid
waste to pass through your anus.
(Up to 24 hours)

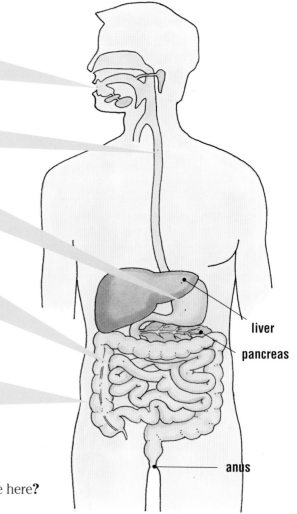

liver

pancreas

anus

d In which part of your gut does food stay the longest**?**
Why do you think this is**?**

e Proteins are digested in your stomach. What are conditions like here**?**

f How long does it take food to pass down the whole length of
your gut**?**

A model gut

You can make a model gut using **Visking tubing**.

1 Wash the Visking tubing in warm water to soften it.
2 Tie one end in a tight knot.
3 Use a syringe to fill the tubing with 5 cm^3 of starch solution and 5 cm^3 of glucose solution.
4 Wash the outside of the tubing.
5 Support your model gut in a boiling tube with an elastic band.
6 Fill the boiling tube with water and leave for 15 minutes.
7 After 15 minutes, test the water for starch and for sugar.

eye protection

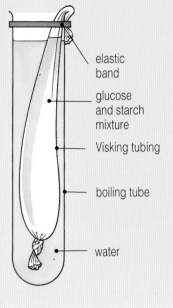

elastic band

glucose and starch mixture

Visking tubing

boiling tube

water

g Which food passed through the tubing into the water?
h Which food did not pass through the tubing into the water?
i Which part of the apparatus was like: i) food in your gut?
ii) your gut wall? iii) the blood around your gut?

Getting through

j How do you think food molecules pass through the Visking tubing?
k Which food molecule (**A** or **B**) is starch and which one is glucose?
l Which food molecule (**A** or **B**) will pass through the Visking tubing?
m Which food molecule, starch or glucose, will pass easily through the wall of your small intestine?
n What would have to happen to food molecule **B** before it could pass through the Visking tubing?
o How do you think this could happen?

mixture of starch and glucose inside

Enzyme investigation

Biological washing powders contain enzymes that can break down food stains.
What might affect how well the enzymes work?
Choose one factor to investigate.
How will you make your investigation safe?
Ask your teacher to check your plans before you start practical work.

allergy

---

1 Match the parts of the body in the first column with the descriptions in the second column:

a) stomach   i) most water is absorbed here.
b) small intestine   ii) saliva is made here.
c) large intestine   iii) most food is absorbed here.
d) mouth   iv) carries food down to the stomach.
e) gullet   v) is very acidic.

2 Find out how each of the following parts of the body help digestion to take place:
a) liver   b) pancreas   c) appendix.

3 Find out the causes of each disease of the gut shown below:
a) constipation   b) stomach ulcers
c) diarrhoea.

Things to do

Questions

1 The graph shows the activity of 2 enzymes: amylase and pepsin.
a) At which pH does pepsin work best?
b) Pepsin is made in the stomach. How does the stomach keep its pH just right for pepsin to work best?
c) Pepsin breaks down proteins. What will be formed as a result?
d) At what pH does amylase work best?
e) Amylase breaks down starch. What will be formed as a result?
f) Amylase works in your mouth. Why is it useful that saliva contains an alkali?

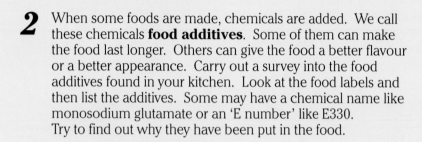

2 When some foods are made, chemicals are added. We call these chemicals **food additives**. Some of them can make the food last longer. Others can give the food a better flavour or a better appearance. Carry out a survey into the food additives found in your kitchen. Look at the food labels and then list the additives. Some may have a chemical name like monosodium glutamate or an 'E number' like E330. Try to find out why they have been put in the food.

3 Plan an investigation to compare the amount of water in a piece of plant food with the amount of water in a piece of meat. You can use the sort of apparatus found in your science laboratory. Remember to make it a fair test and check your plan with your teacher before carrying it out.

4 Visit your local supermarket. Find out the cost of different food groups using the examples in the table:
a) Which food can you buy most of for £5?
b) Which food can you buy least of for £5?

Some families have more to spend on food than others.
c) Which food groups do you think a family on a very low income would have to buy, to feed themselves?
d) Which food groups would a high-income family be able to buy?
e) Do you think your answers to c) and d) would be true in India and in the United States?

Food group	Example	Cost per 1 kg
carbohydrates	potatoes rice	
fats	cheese butter	
proteins	chicken lamb	
vitamins and minerals	oranges broccoli	

5 Your teacher will give you a table of Recommended Daily Amounts of nutrients (RDAs).
a) Write down how much energy you need. How does this compare with:
  i) a 1 year old?
  ii) someone the same age as you but of the opposite sex?
  iii) a very active adult female?
b) Which group shows the biggest increase in protein needs for each sex? Why do you think this is?
c) Which foods does a pregnant woman need more of than a female desk worker? Try to explain these differences.

6 Plan the following meals, choosing foods that would be good for you:
a) A good breakfast for a 12 year old.
b) A high-energy lunch for an athlete before a big race.
c) An evening meal low in fat but rich in fibre and protein.

Respiration

How long could you live without oxygen? Not long.
Your lungs take oxygen out of the air.
In your lungs the oxygen passes into your blood.
Your heart pumps the blood all round your body to your
cells: and that's where respiration takes place.

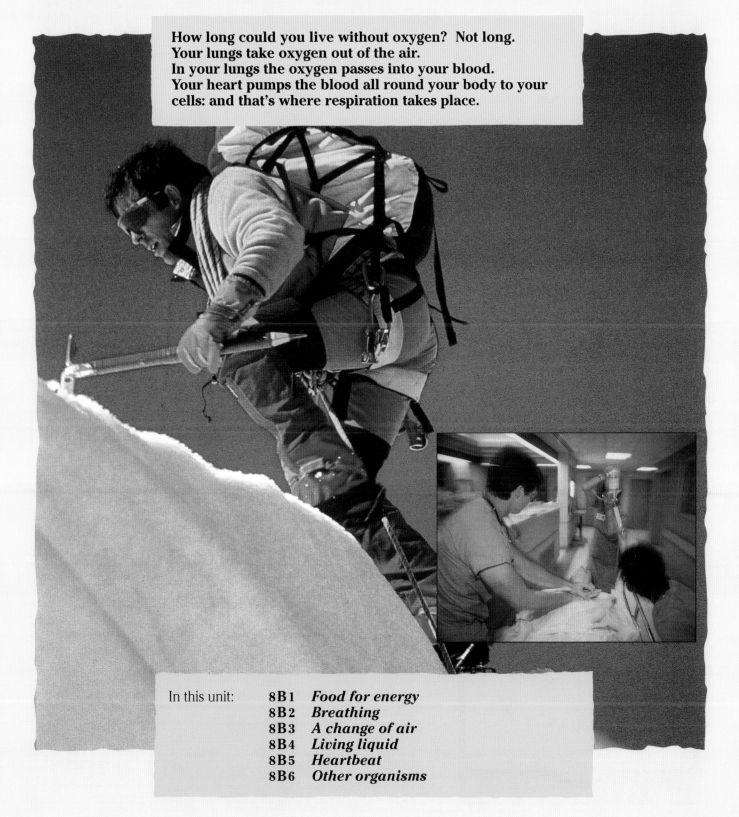

13

Food for energy

You need energy for running, sitting, breathing and even for sleeping.
In fact, everything you do needs energy.

You get your energy from your food.
The food you eat is your fuel.
Almost everything you eat contains energy.

Energy in food is measured in **kilojoules (kJ)**, where **1 kilojoule = 1000 joules**.

cornflakes with milk
700 kJ

yoghurt
400 kJ

sausage roll
1500 kJ

chips
1000 kJ

tea, milk and sugar
200 kJ

▶ Look at the potential energy in these foods:

a How much energy is there in a breakfast of cornflakes, yoghurt and a cup of tea?

b How much energy is there in a meal of chips and 2 sausage rolls?

'High energy' drinks that athletes use contain a lot of glucose.
This provides energy directly for their cells to use.

How much energy?

The energy stored in foods is often shown on the label.
It is usually shown in kilojoules (**kJ**), and also in kilocalories (kcal).
(kcal is a unit of energy often used in slimming diets.)

▶ Look at these food labels.

c Why is the energy given for 100 grams of each food?

d Which of these foods has the lowest energy per 100 g?

e What happens if the food you eat contains more energy than you need?

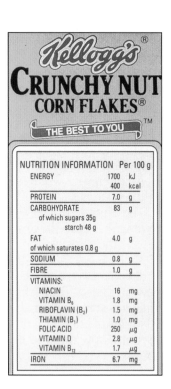

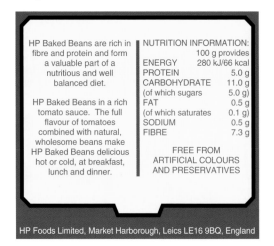

HP Baked Beans are rich in fibre and protein and form a valuable part of a nutritious and well balanced diet.

HP Baked Beans in a rich tomato sauce. The full flavour of tomatoes combined with natural, wholesome beans make HP Baked Beans delicious hot or cold, at breakfast, lunch and dinner.

NUTRITION INFORMATION:	
	100 g provides
ENERGY	280 kJ/66 kcal
PROTEIN	5.0 g
CARBOHYDRATE	11.0 g
(of which sugars	5.0 g)
FAT	0.5 g
(of which saturates	0.1 g)
SODIUM	0.5 g
FIBRE	7.3 g

FREE FROM
ARTIFICIAL COLOURS
AND PRESERVATIVES

HP Foods Limited, Market Harborough, Leics LE16 9BQ, England

NUTRITION
Sainsbury's Sardines in Brine are a good source of Calcium and Vitamin D, both needed for strong bones and teeth; Vitamin B$_{12}$, required for healthy blood and nervous system, Niacin which helps food to give us energy.

	TYPICAL VALUES PER 100 g (3½ oz) OF DRAINED PRODUCT
ENERGY	170 kCALORIES 705 kJOULES
PROTEIN	23.4 g
CARBOHYDRATE	less than 0.1 g
TOTAL FAT	8.3 g
ADDED SALT	0.5 g

VITAMINS/ MINERALS	% OF RECOMMENDED DAILY AMOUNT
NIACIN	45%
VITAMIN B$_{12}$	1400%
VITAMIN D	300%
CALCIUM	110%
IRON	25%

Kellogg's
CRUNCHY NUT
CORN FLAKES®
THE BEST TO YOU ™

NUTRITION INFORMATION	Per 100 g	
ENERGY	1700	kJ
	400	kcal
PROTEIN	7.0	g
CARBOHYDRATE	83	g
of which sugars 35g		
starch 48 g		
FAT	4.0	g
of which saturates 0.8 g		
SODIUM	0.8	g
FIBRE	1.0	g
VITAMINS:		
NIACIN	16	mg
VITAMIN B$_6$	1.8	mg
RIBOFLAVIN (B$_2$)	1.5	mg
THIAMIN (B$_1$)	1.0	mg
FOLIC ACID	250	μg
VITAMIN D	2.8	μg
VITAMIN B$_{12}$	1.7	μg
IRON	6.7	mg

Investigating the energy in food

One way of measuring the amount of energy in some food is to burn it.

As the food burns, it gives out energy. We can use this energy to heat up some water.

The more energy stored in the food, the more energy is released and the hotter the water gets.

Plan an investigation to compare the energy content of a peanut with that of a pea.

- What apparatus will you need?
- What measurements will you take?
- How will you record your results?

Remember you must make it a **fair test**, and work safely.

When you have had your plan checked by your teacher, go ahead and do the investigation.

What do you find?

The energy stored in food is released in our cells in the process called **respiration** (see lesson 8B2). What differences do you think there are between the reaction shown above and the reactions in our cells?

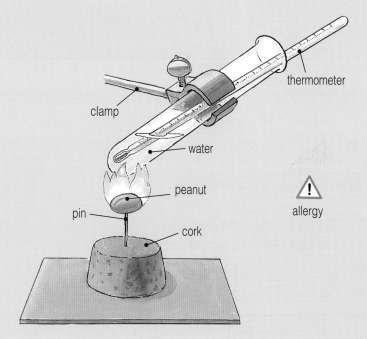

clamp

thermometer

water

peanut

pin

cork

allergy

---

1 Proteins, carbohydrates, vitamins and fats are all food groups. Put these 4 groups in order starting with the one that gives the most energy per gram and ending with the one that gives the least.

2 Make a survey of how much energy is in different foods. Look at the food labels on packets and cans at home. List them under 'high energy food' or 'low energy food'.

3 Look at the bar-chart.
It shows the energy content in kilojoules for one gram of each food.
a) Which food gives the most energy?
b) Which two foods give the least energy?
c) Which foods would you take with you on a long walk in the mountains?
d) Which foods would make a good meal for someone who wants to lose weight?
e) How much energy would you get from 1 gram of bread?
f) How much energy would you get from 2 grams of carrot?

4 In your investigation, did all the energy from the peanut go to the water?
Was it a fair test? What could you do to improve your investigation?

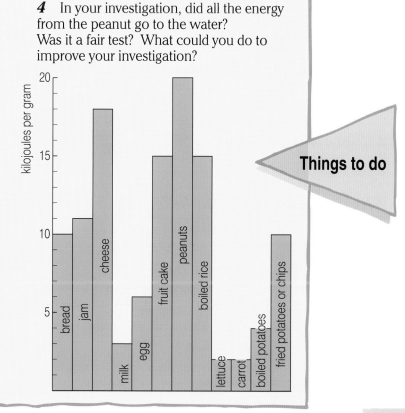

Things to do

Breathing

We all exercise at some time. What exercise have you had in the last week**?**

Do you feel different after exercise**?**

▶ Write down some of the changes that happen to your body when you exercise.

Puffing and panting

1 Sit still and count how many times you breathe out in 1 minute.
This is your rate of breathing at rest.

2 Fasten a breathing sensor around your chest.
This will not only measure your ***breathing rate*** but will also give
you an idea of how ***deep*** each breath is.

3 Start logging your breathing movements.

4 Copy this table and fill in your breathing rate at rest (you can work
this out by counting the number of peaks on the graph).

Breathing rate at rest (breaths per minute)	Breathing rate after light exercise (breaths per minute)	Breathing rate after heavy exercise (breaths per minute)

Make sure the box or bench you step onto is firmly in place.

5 Now do step-ups for 1 minute (light exercise).
As soon as you have finished, sit down and measure your
breathing rate.
Put your reading in the table.

6 Now do step-ups as quickly as you can for 3 minutes (heavy exercise).
As soon as you have finished, sit down and measure your breathing rate.
Put your reading in the table.

7 Look at the bar-chart data for your breathing rates.

a What happened to   i) your breathing rate in the investigation**?**
ii) the size of each breath in the investigation**?**
b Why do you think this happened**?**
c What do you think affects your breathing rate and your depth of breathing**?**

Why do we have to breathe?

All the cells in your body need energy to stay alive.
Can you remember where you get your energy**?**

You get energy out of your food in **respiration**:

SUGAR + OXYGEN → CARBON DIOXIDE + WATER + ENERGY

Oxygen is needed for respiration to happen.
When sugar is burnt in oxygen, it gives out energy for your cells to use.
You get oxygen into your body by breathing it in.

▶ Use this information to explain why your breathing rate went up
when you did more exercise.

How do you get oxygen into your body?

When you breathe in, air goes down your wind-pipe to your lungs.
Each lung is about the size of a rugby ball.

d Where do you think your lungs are found?

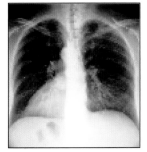
A chest X-ray

Look at the 2 diagrams.
The one at the top shows what the inside of your chest looks like.
The one at the bottom shows a chest model.

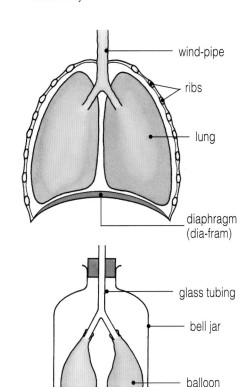

e Which part of the human chest is shown by the i) balloons?
ii) bell-jar?   iii) rubber sheet?   iv) glass tube?

f When you breathe in, do your lungs get bigger (inflate) or
smaller (deflate)?

g What happens to your lungs when you breathe out?

▶ Get the chest model.
Pull the rubber sheet down and then push it up.
Do this a few more times and watch what happens to the balloons.

h When you breathe in, does your **diaphragm** move up or down?

i What happens to it when you breathe out?

▶ Measure the size of your chest with a tape.
Now take in a deep breath.

j What happens to the size of your chest when you breathe in?

k What happens when you breathe out?

▶ Put your hands on your chest.
Breathe in and out deeply and slowly.

l Which way do your ribs move when you breathe in and out?

Muscles raise and lower your ribs and raise and lower your diaphragm.

Things to do

1 Copy and complete the table using the information on this page.

	Breathing in	Breathing out
What do the ribs do? What does the diaphragm do? What happens to the space inside your chest? What happens to your lungs?		

2 Try making your own 'model lungs'.
You could use an old plastic lemonade
bottle, balloons, a rubber-band and a plastic
drinking straw.
Make a hole in the top for the straw and
make it air-tight with plasticine. Cut away
the bottom of the bottle and stretch a
balloon over it for a diaphragm.

3 At the top of high mountains there is far
less oxygen in the air.
a) How do mountaineers manage to
   breathe?
b) Why do you think many athletes train at
   high altitude?

A change of air

Learn about:
- inhaled and exhaled air
- how our lungs work

"You breathe in oxygen and breathe out carbon dioxide" said Robert. Do you think that he is right?

▶ Try breathing out through a straw into a test-tube of lime water. What change did you see? This is a test for carbon dioxide.

⚠ Care! don't suck. Dispose of the straw in a waste bin.

▶ Look at the pie-charts:

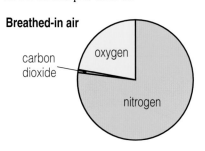

Breathed-in air

carbon dioxide

oxygen

nitrogen

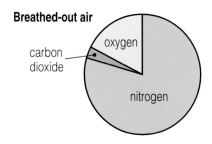

Breathed-out air

carbon dioxide

oxygen

nitrogen

a Which gas makes up most of the air you breathe in?

b Which gas do you breathe in more of?

c Which gas do you breathe out more of?

eye protection

▶ See what happens to a candle when it is left to burn in i) fresh air and ii) breathed-out air (your teacher will show you how to collect this). (Hint: a stop-watch might be helpful.)

d What did you observe?

e Try to explain what you saw.

In and out ...

Set up the apparatus as shown in the diagram:

Breathe gently in and out of the mouth-piece several times.

f When you breathe in, does the air come in through A or through B?

g When you breathe out, does the air go out through A or through B?

h In which tube did the lime water turn cloudy first?

Write down your conclusions for your experiment.

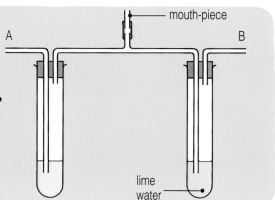

mouth-piece

A   B

lime water

Gail said "You must have the same volume of lime water in each tube". Can you explain why she is right?

Look back at what Robert said at the top of the page. Can you write out a better sentence?

Looking into your lungs

The diagram shows you how air gets to your lungs through your wind-pipe and then through the air passages.
The air passages end in tiny bags called **alveoli (air sacs)**.
These have very thin walls.
They are surrounded by lots of tiny **blood vessels**.

i How do you think oxygen gets from your lungs to all the cells of your body?

j How do you think carbon dioxide gets from the cells of your body to your lungs?

k Which gas do you think passes from the alveoli into the blood vessels?

l Which gas do you think passes in the opposite direction?

m These gases are swopped very quickly.
Write down 2 things that help this to happen (hint: read the sentences above again).

n What do you think will happen to the process in **m** if the lungs are damaged?

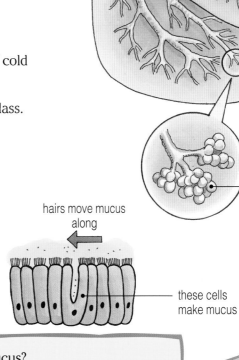

A lot of hot air

▶ Try breathing out onto a cold glass surface, like a beaker of cold water or a window.
What do you see?
Now put a strip of blue **cobalt chloride paper** onto the glass.
(Rinse your hands after handling the paper.)
What happens?
Write down your conclusions.

Can you think of any other differences between the air you breathe in and the air you breathe out?
For one thing, the air you breathe out is *cleaner*.
Your air passages are lined with a slimy liquid called **mucus**.
This traps dust and germs.
Then millions of microscopic **hairs** carry the mucus up to your nose and throat.

hairs move mucus along

these cells make mucus

---

1 Copy and complete:
We breathe in air containing nitrogen, ... and some carbon dioxide. The air that we breathe . . . . contains the same amount of . . . ., less . . . . and . . . . carbon dioxide. The air we breathe out also contains more . . . . vapour and is at a . . . . temperature.

2 Do you know about **artificial respiration**?
It's a way of starting up someone's breathing again.
You can learn about it at a first-aid class.
Your teacher can give you a Help Sheet explaining how it works.

3 What is mucus?
How does it help to clean up the air that you breathe in?

4 When you breathe in you take fresh air into your lungs.
a) Write down the pathway taken by oxygen from the air outside until it enters your blood.
b) What happens to the volume of air you breathe in when you exercise?

Things to do

Living liquid

Learn about:
- cells needing oxygen
- transport in blood
- veins and arteries

What does blood make you think of?
Horror movies, vampires, wars?

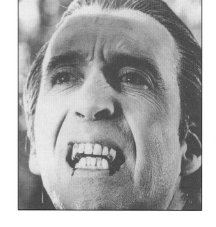

You probably have about 4 litres of blood in your body.
That's a bucket-full.
It's flowing around your body all the time.
But what is it for?

▶ Write down your ideas about how your blood helps you.

You already know that the cells of your body need food and oxygen
to give you energy in respiration.
Your cells make waste chemicals too.
Your kidneys get rid of most of these waste chemicals.

▶ Look at the diagram and answer the questions:

HEART

f What does the blood
drop off at the kidneys?

KIDNEYS

g What keeps
the blood
moving round
and round?

Start here
a What does the blood
collect when it is here?

INTESTINES

e What does the blood take
away from the cells?

d Name 2 things that the blood
drops off that the cells use to live.

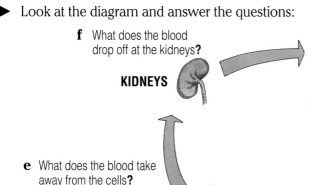

**CELLS IN
THE BODY**

LUNGS

b What does the blood
collect when it is here?

c What does the blood
drop off here?

Return journey

So your blood carries many things around your body.
It is rather like a railway system, where trains pick things
up at one place and deliver them to another.

h How do you think the blood is kept on the move?

The blood is carried around your body in tubes called **blood vessels**.

▶ Look at the diagram:

i What do we call tubes that carry blood away from the heart?

j What do we call tubes that carry blood back to the heart?

With your finger, trace the path taken by the blood from:
- the heart to the body
- the body to the heart
- the heart to the lungs
- the lungs to the heart.

Near to the cells are the tiniest blood vessels called **capillaries**.
These connect your arteries to your veins.

k Why do you think that capillaries have very thin walls?

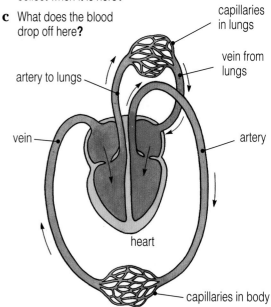

capillaries
in lungs

vein from
lungs

artery to lungs

vein

artery

heart

capillaries in body

Around and around ...

Do you know how to take your pulse?
When you take your pulse you feel an artery.
Blood flows through your arteries in spurts.
This is the pulse that you feel.

Does your pulse rate and breathing rate go up and down together?

Plan an investigation to see how your pulse and breathing are affected by exercise.

- What exercise will you plan to do?
- What measurements will you take?
- How will you make it a fair and safe test?
- How do you plan to show your results?
- Check your plan with your teacher before carrying it out.

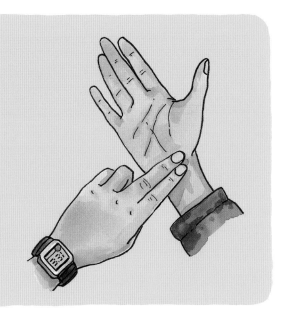

Supply lines

▶ Lift one arm above your head and let the other arm hang at your side.
Keep them there for a minute or two.
Now bring them in front of you and look at the differences between the veins on the back of each hand.
What differences can you see?

▶ Look at the photograph of a section of an artery and a vein.

l What differences can you see?

m Can you find out any other differences between arteries and veins?

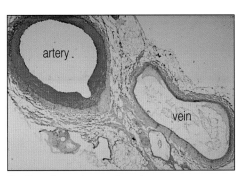

Section of an artery and a vein

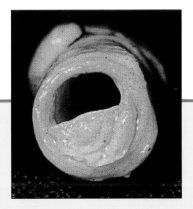

Things to do

1 Copy and complete:
Blood is pumped around my body by my
. . . . Blood travels away from my heart in
. . . . and back to my heart in . . . . The tiniest
blood vessels are called . . . . and these have
very . . . . walls so things can pass in and out.
When I feel my pulse I am touching an . . . .

2 Make a table of the differences between
arteries and veins. Your teacher can give
you a Help Sheet.

3 Sometimes our arteries can get 'furred
up'. This is because a fatty substance sticks
to the inside of the artery and makes it
narrower. How do you think this would
affect the flow of blood in the artery?

Heartbeat

Learn about:
● how the heart works
● heart disease
● early ideas about circulation

What do you think is the strongest muscle in your body?
Not many people think that it is their heart.
Just think of the job your heart does.
It beats about 70 times a minute, for 60 minutes per hour and 24 hours a day, to keep you alive.

a Use a calculator to work out how many times your heart beats:
i) per hour ii) per day iii) in a year.

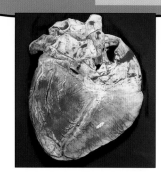
A human heart

The double pump

Where do you think your heart is?
Put your hand on the place where you think it is.
What can you feel?

b How do you think your heart is protected?

c How many spaces are there inside your heart?

▶ Look at the diagram. It is drawn as though you are facing someone.

Your heart is really 2 separate pumps side-by-side.
When your heart beats, the muscle squeezes the blood out.

d Where does the right-side of your heart pump the blood to?

e Where does the left-side of your heart pump the blood to?

f Which side of the heart will have blood containing most oxygen?

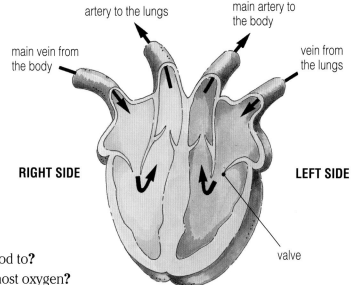

artery to the lungs

main artery to the body

main vein from the body

vein from the lungs

RIGHT SIDE

LEFT SIDE

valve

Heart listening

The 2 pumps both beat at the same time.
You can hear your partner's heartbeat by using a **stethoscope**.
Try making a home-made version as shown in the diagram.
What sounds can you hear?

You should hear 2 sounds.
Doctors call them lub-dub sounds.
The 2 noises are caused by the **valves** in your heart closing.
Try listening again . . . lub-dub . . . lub-dub . . . lub-dub . . .

g Why do you think your heart valves close?
(Hint: look at the top diagram.)

Find out if your heartbeat (the number of beats per minute) is the same as your pulse rate.

h Does it change in the same way as your pulse rate when you exercise?

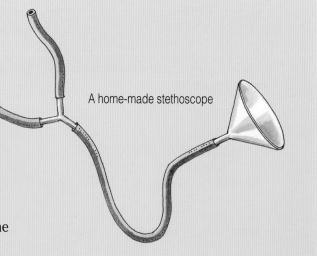
A home-made stethoscope

The big killer

Heart disease is one of the biggest killers in Britain.
Fatty substances can 'fur up' the arteries leading to the heart muscle.

i What would 'furring up' do to the flow of blood to the heart muscle?

j What could happen to the supply of oxygen to the heart muscle?

If the heart muscle does not get enough oxygen it can cause chest pains.
This is called **angina**.
It is a warning that the person is more likely to have a heart attack.

Sometimes a clot can form inside a **coronary artery**.

k How do you think this could cause a **coronary heart attack?**

▶ Look at the cartoons:
In your groups discuss the things that you think can increase the risk of heart attack.
How do you think each of these risks could be reduced?
Make a list of your ideas.

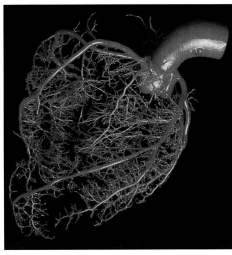

A cast of the coronary arteries

---

1 Copy and complete:
The heart is made out of . . . . The blood on the left-side contains . . . . oxygen than the blood on the . . . . -side. This is because the blood has just come back from the . . . . The left-side of the heart pumps blood all around the . . . . The heart has . . . . to stop the blood from flowing backwards.

2 Look at the heart diagram opposite.
List what happens to the blood as it passes from the main vein to the heart and eventually to the main artery.

3 What sort of person do you think is most likely to suffer a heart attack? How old will they be? What is their weight like? What sort of habits might they have?
Draw a cartoon of the person and label it.

4 Design a leaflet or make a poster to let people know about the risks of heart disease.

5 Choose one of the following scientists to research: Galen; Vesalius; Harvey; Withering; Ibn-al-Nafis.
How and what did they find out about the heart and circulation?

Things to do

Other organisms

Learn about:
● respiration in plants
● factors affecting respiration

▶ Make a list of some of the living things that you have studied during your Spotlight Science course.

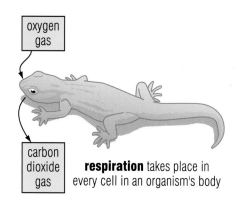

oxygen gas

carbon dioxide gas

respiration takes place in every cell in an organism's body

All of these living things including animals, plants, moulds and bacteria are able to carry out **respiration**.

Do you remember the word equation for respiration?

SUGAR + OXYGEN → CARBON DIOXIDE + WATER + ENERGY

How fast?

How could you find out how quickly respiration is going?
Well, the faster the rate of respiration, the more carbon dioxide is produced.

Can you remember how you showed that carbon dioxide is produced during respiration?

Carbon dioxide also turns red hydrogencarbonate indicator yellow.

Therefore the quicker the indicator turns yellow, the faster respiration is taking place.
You will need to remember this when you do your investigation.

Respiration in plants

All plants carry out respiration to get energy.

a During respiration, how do you think oxygen gets into a plant and carbon dioxide gets out?

b During photosynthesis:
  i) which gas passes into a leaf? and
  ii) which gas passes out of a leaf?

Some people think that plants carry out photosynthesis during the day and respiration at night. This is wrong.

c Look at the diagram and say what really happens.

As you can see, plants carry out respiration **all the time**.
But they carry out photosynthesis **only in the light**.

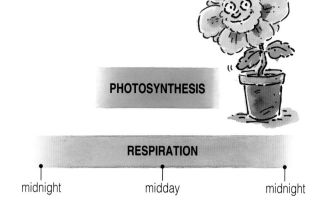

PHOTOSYNTHESIS

RESPIRATION

midnight   midday   midnight

More or less carbon dioxide?

As you know, the faster respiration takes place, the more carbon dioxide will be produced in a given time.

Plan an investigation to find out what factors affect the rate at which carbon dioxide is produced.

You can use the simple apparatus shown in the diagram:

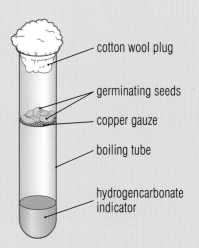

- Choose from a selection of living things: yeast, germinating seeds, woodlice or maggots.

- Think about what factors will affect your enquiry:
  – How much living material will you use?
  – What will the volume and concentration of indicator be?
  – At what temperature, and for how long, will you leave the apparatus?

- Set up a control apparatus (without living material in it).

- Show your plan to your teacher and then do it. Your title should be the question that you have chosen to investigate.

Be careful not to damage any small animals that you use.

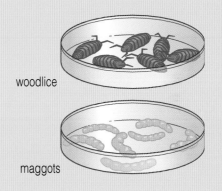

- After you have completed your enquiry, compare your results with those of other groups.

- Try to identify any trends in the data you collected.

1 Copy and complete:
All living things carry out . . . . The faster the rate of respiration, the more . . . . is produced. We can test for carbon dioxide with . . . . indicator. It will turn from red to . . . . Active animals produce . . . . carbon dioxide. Plants usually make . . . . carbon dioxide than animals. Plants carry out . . . . all the time and photosynthesis only in the . . . .

2 Try to explain how the following might affect the amount of carbon dioxide made in respiration: a) exercise b) temperature c) using plant or animal material.

3 a) If you were to monitor the oxygen and carbon dioxide levels in an aquarium over 24 hours, how do you think they would vary?
b) What would happen if you were to put more plants into the water?
c) What would happen if you were to put more animals into the water?

4 Look back at the word equation for respiration. List the ways in which you could measure the rate of respiration in germinating seeds.

Things to do

Questions

1. Gemma and Beth measured their breathing rate (in breaths per minute) before they ran a race. Then they measured their breathing rates again after the race, every minute, until their rates returned to normal. They recorded their results in a table:

	Before exercise	Minutes after exercise						
		1	2	3	4	5	6	7
Gemma	16	45	38	31	24	20	17	16
Beth	13	35	32	28	22	18	13	13

   a) Plot 2 line-graphs on the same sheet. Use the vertical axis for breathing rate and the horizontal axis for time.
   b) Who took the longer time to recover from the exercise?
   c) Who do you think was the fitter of the two girls? Give your reasons.

2. Where in your body are the following found:
   a) diaphragm?   b) alveoli?   c) valves?   d) capillaries?
   What job does each one do?

3. Do you think the following statements are true or false?
   a) Most of the air you breathe is not used by your body.
   b) Carbon monoxide is a waste product of respiration.
   c) Respiration produces oxygen gas.
   d) Exercise increases the risk of heart disease.

4. Coronary heart disease can be caused by eating too much saturated fat in foods like meat, butter and cream. You should choose unsaturated fats, in foods like fish and vegetable oils.
   Look at food packets and make a list of foods for each of these 2 types of fat.

5. Blood has many different jobs.
   Which part of your blood:
   a) carries oxygen?   b) carries dissolved food?
   c) fights germs?   d) helps your blood to clot?

6. a) Find out where your red blood cells are made.
   b) Find out how lack of iron causes **anaemia** and how this affects the body.
   c) Why do you think that women take more iron tablets than men?

7. Look at this blood transfusion card:
   Can you name the 4 main blood groups?
   Make a leaflet or a poster that will persuade people to give blood.

I do something amazing.
I give blood.

BLOOD GROUP   DONOR NO.

Microbes and disease

We are surrounded by different types of microbes all the time.

Some of these living things are very useful to us, but others are potentially harmful.

In this unit you will find out how your body can defend you from harmful microbes and how scientists have helped to protect us against some diseases.

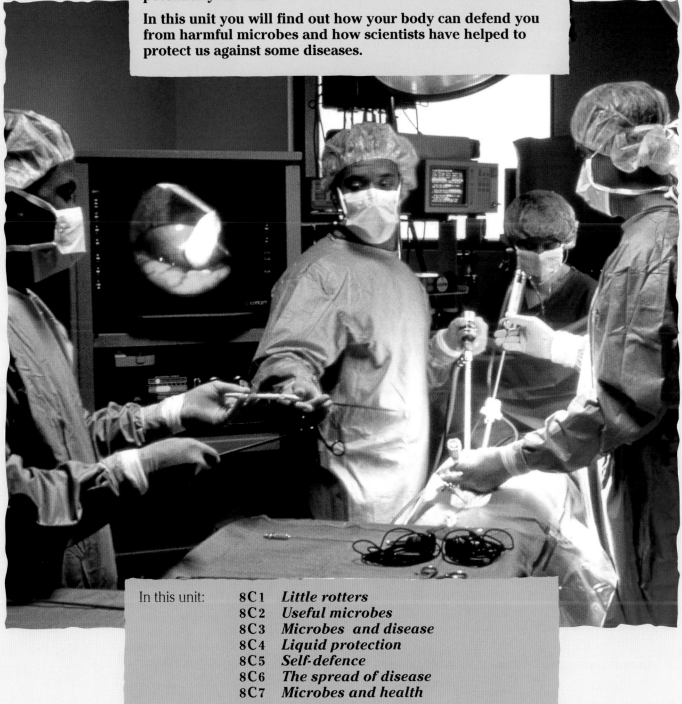

Look at this photograph of food that has gone off. In olden days people did not know why food went bad. Now we know that **microbes** are to blame.

Microbes are all around us. They are in the air we breathe, in the soil and in untreated water.

Microbes include **bacteria**, **fungi** (moulds) and **viruses**. Many are so small that you need a microscope to see them properly.

Mould growing on some nectarines

Looking at moulds

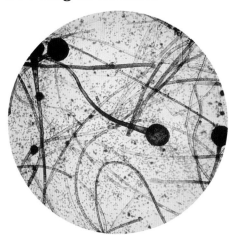

In a fume cupboard, using tweezers place a small piece of bread mould on to a slide.

Add a drop of water and gently lower a cover slip on to the mould. If it does not lie flat, gently tap the cover slip with the end of your tweezers.

▶ Look at the bread mould under your microscope.
Describe exactly what you see.

Bread mould under the microscope

Picking some mould off old bread with tweezers

allergy

Lowering cover slip onto mould on a slide

Mouldy bread!

What affects the growth of microbes?

Old bread often goes mouldy.
How could you find out what affects how quickly bread goes mouldy?

Plan an investigation.

How would you make it a fair test?
Decide how you would record your results.
Can you predict any patterns that your results might have?

▶ Which conditions do you think *stop* the following foods from going bad?

a frozen sweetcorn.   **d** canned peaches.

b dried peas.   **e** packet of crisps.

c vacuum-packed bacon.

Something in the air?

It is now known that microbes make food go bad, but how do they reach the food in the first place?

In this experiment you are going to use **nutrient broth** to grow microbes.

In order to make it a fair test the broth must be **sterilised**. This means that any microbes already present will be killed.

Take 4 test-tubes and pour in nutrient broth to fill each to about a third full.

Set them up as shown in this diagram.

Get your teacher to heat tubes A, B, C and D to a high temperature in a pressure cooker for 15 minutes.

What do you think this will do to the nutrient broth?

Label your test-tubes. Place them in a beaker and leave them at room temperature for one week.

Copy out the table so that you are ready to record your results.

If microbes are present they will turn the broth cloudy.

f In which test-tubes did the broth turn cloudy?

g How did the microbes get to the broth in these tubes?

h Do you think microbes are lighter or heavier than air? Give your reason.

i Why did microbes not get into the broth in test-tube D?

Tube	A	B	C	D
Clear or cloudy?				

---

1 The Cook family went away on their summer holiday for 3 weeks. They left some food out in the kitchen in their hurry to leave.
What do you think will have happened to each of the following foods by the time they return?

packet of cornflakes   open bottle of milk
bowl of sugar   jar of jam (with lid off)
piece of cheese   tin of grapefruit
apple   packet of salted peanuts
bowl of cat meat

2 Find out about 3 microbes that are harmful to people and 3 microbes that are useful.

3 Many food packets have 'sell by' or 'use by' dates on them.
Make a list of some of the foods that are marked with these dates.
What do you think might happen to the food after this date?
Canned and dried foods don't need a 'sell by' date. Why is this?

4 Louis Pasteur was a French scientist famous for his work on microbes. Find out about his work using books, ROMs or the internet.

Things to do

8C2 Useful microbes

Learn about:
- fermentation
- investigating yeast

You get energy out of your food in respiration.

a Complete the word equation for respiration:

SUGAR + OXYGEN ⟶ .... + ... + ENERGY

Respiration is an **exothermic** reaction.

b What does exothermic mean? (See page 80 in Book 7.)

c Think about respiration and burning.
What are the similarities between these reactions?
Make a list of your ideas.

d Why do you get short of breath and hot in a race?

Fermentation

When **we** respire we use oxygen. It reacts with our food to make carbon dioxide, water and energy.
But some living things can respire **without** oxygen.
An example is a microbe called **yeast**. Yeast is a type of fungus.
We can use yeast to make wine.
It uses up the sugar in fruit to make alcohol.
This process is called **fermentation**.

Bubbling alcohol

Try your own fermentation using sugar.
Mix some yeast with about 4 g of sugar.
Add about 10 cm³ of warm water (about 35°C) and put it in a flask.
Leave the apparatus set up like the diagram shows.
Look at the flask after 10 minutes. What do you notice?
Look at the flask again after 45 minutes.

⚠ **Carefully** smell the contents. What do you notice?

⚠ **Do** not **attempt to drink this substance. It is very impure.**

Plan a safe investigation to find out what affects the rate of yeast fermentation.
Work with other groups to investigate different factors.
Show your plan to your teacher before you start.

lime water

sugar + yeast + water

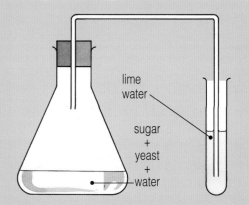

The word equation for fermentation is

| sugar (with yeast) | ⟶ alcohol + carbon dioxide + energy |

e In what ways is this reaction **like** your respiration?

f In what ways is this reaction **different** to your respiration?

A fermenting mixture in a brewery

Other uses of microbes

Microbes have many uses, besides making alcohol.
Look at the photographs below:

yoghurt

bread

cheese

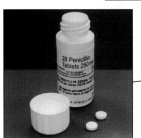
making antibiotics

useful microbes

meat substitute,
e.g. Quorn

treating sewage

biofuels

Things to do

1 Copy and complete:
a) When we breathe, the products are . . . . , ... and . . . .
b) The products of fermentation are . . . . , ... and . . . .
c) An exothermic reaction is one in which heat . . . .

2 Yeast is also used in bread-making.
Find out how bread is made.
Why is yeast used?
Why could this be called a ***bubbling reaction***?

3 Remind yourself of the 'Bubbling alcohol' experiment opposite.
How do you know a reaction has taken place? List your ideas.

4 Breath tests can be used to see if drivers have been drinking too much alcohol.

a) Make a list of arguments ***for*** random breath-testing (testing any driver at any time).
b) Make a list of arguments ***against*** random breath-testing.

8C3

Microbes and disease

Learn about:
- causes of disease
- how microbes enter the body
- researching disease

Do you know why people get ill**?**

▶ Make a list of some of the things you think make people ill.

Your skin acts as a barrier to microbes.

a Can you remember the main types of microbes**?**

Bacteria and viruses are a common cause of disease.

b Make a list of some diseases caused by these microbes.

The **symptoms** of a disease are the body's response to waste chemicals made by the microbes.

c Write down some symptoms that you know of.

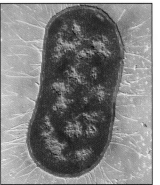

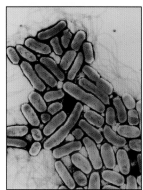

Points of entry

Look at these ways in which diseases can be spread:

     air   food   touch   water   animals

d For each method shown in the picture, write down:
   i) A disease that can be spread in this way (your teacher can give you a Help Sheet).
   ii) How its spread can be prevented.

e Can you think of any other ways in which diseases can be spread**?**

▶ Use your Help Sheet of diseases to find out the answers to the following questions.

f What type of microbe causes   i) tuberculosis**?**   ii) measles**?**   iii) athlete's foot**?**

g What are the symptoms of   i) polio**?**   ii) mumps**?**   iii) the common cold**?**

h How are   i) malaria   ii) rubella, and   iii) athlete's foot spread**?**

Seldomill Health Authority

Memo to: *Analysts*   **From:** *Mike Robe*

The children at Sick Lee High School have been going down with severe stomach upsets. I think that the disease may be linked to places where they eat their lunch. These are the school canteen, the Greasy Cod Chip Shop, Sid's Snack Bar and Betty's Bakery.
Please plan an investigation to find out the source of infection. Write me a report about your plan, the tests you intend to use, and how you will show your results.

Please hurry!

Antibiotics: useful drugs

Your doctor might give you an **antibiotic** to help you fight a disease.
Antibiotics kill some bacteria or stop them from growing.
However, they do not work against a virus (so they can't cure a cold!).
The first antibiotic was discovered by Alexander Fleming.

Research activity

In this enquiry you can find out how people in the past fought against the spread of disease.

Make a poster to share your findings with other groups.

Choose one of these topics:

* the plague in Eyam, Derbyshire
* cholera and Dr John Snow
* yellow fever and Dr Carlos Finlay.

Just like Fleming!

* Using sterile forceps, place a sterile paper disc into each of:

   A – disinfectant   **B** – alcohol   **C** – crystal violet   **D** – washing-up liquid

* Leave to soak for 5 minutes.
* You will be given an agar plate which has harmless bacteria growing on it.
* Divide the underneath of an agar plate into 6 sections and label them **A** to **F**.
* Remove the discs with sterile forceps and shake off any excess liquid.
* Place each disc on the correct part of the agar plate.
* Your teacher will give you a **penicillin** and a **streptomycin** disc for sectors **E** and **F**.
* Sellotape the lid to the base so it cannot come off (see photo):
* Incubate your plate for 48 hours at 25°C. Do not open plate. Then examine the growth of bacteria.

⚠ do not open plate after incubation

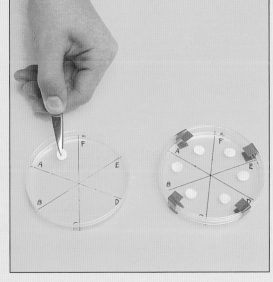

i   Which chemical had most effect upon the growth of the bacteria?

j   How were you able to measure this effect?

k   How could you use this test to find out the best concentration of an antibiotic to use?

Things to do

1   Copy and complete:
Bacteria and . . . . are the main microbes that cause disease. The . . . . of a disease are caused by chemicals made by the microbes. A drug that fights the disease inside your body is called an . . . . Alexander . . . . discovered one of the first antibiotics.

2   Name 4 ways in which diseases can be spread.
For each way say how you think the disease can be prevented.

3   Drugs that kill microbes inside the body are called **antibiotics**.
What are **antiseptics** and **disinfectants**? How do they help to fight disease?

4   How do you think the following can help to spread disease:
a) flies?   b) hypodermic needles?
c) kitchen cloths?
d) Make a leaflet advising 'back packers' on how to avoid infections on a trip to a remote location.

Liquid protection

Learn about:
● the role of blood
● antibodies in blood

Do you know what a **blood transfusion** is?
It once saved Richard's life.
He was involved in a motorway accident.
He was losing a lot of blood.
Fortunately for him, the ambulance team
were quick to arrive.
So they were able to give him extra blood.
But not all our blood is the same.
Richard carried a card that showed which
type his blood was.

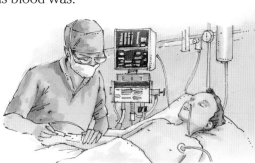

a There are 4 main blood types. Do you know what they are called?

Each blood type contains slightly different chemicals.

b Why do you think that you can only have a transfusion of blood of the same type?

Richard was very grateful to the blood donors who gave their blood.

c What do you have to do to be a blood donor?

Blood bank

Blood taken from a donor is treated so that it does not clot.
It may settle out into 2 parts. A pale yellow liquid called **plasma**
and a deep red layer of **blood cells**.

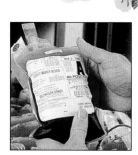

d Is the blood made up mainly of plasma or blood cells?

Plasma is mainly water containing dissolved chemicals.

Your teacher will give you a prepared slide of blood cells.
Look at it under your microscope.
You will need to focus carefully at high power.

e Which do you think are bigger: red cells or white cells?

f Are there more red cells or white cells?

Draw a diagram of each different blood cell that you can see.

g Can you see any other differences between red and white cells?

h Why do you think the white cells look purple?

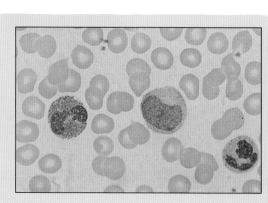

The oxygen carriers

The red cells carry oxygen.
They are red because they contain **haemoglobin**.
This is a chemical that can collect and carry oxygen.
Haemoglobin lets go of oxygen when it comes to a part of the body that needs it.

i Where do you think haemoglobin collects oxygen from**?**

▶ Copy this diagram. Add as many labels and notes to it as you can to explain how oxygen is carried round your body.

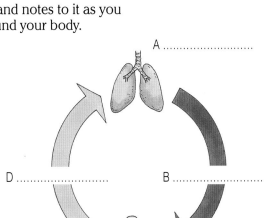

A ....

B ....

C ....

D ....

The protection gang

White cells protect your body from germs.

j In what ways could germs get into your body**?**

One sort of white cell **eats** any germs that it finds.
Another sort makes chemicals (called **antibodies**) that can kill germs.

k Can you see 2 types of white cell in the photograph at the bottom of page 34**?**
In what ways do they look different**?**

l What happens just after you cut yourself**?**

Eventually a scab will form.
But first the bleeding has to be stopped.
The cut is sealed by lots of tiny bits of cells called **platelets**.

m Why is it important that the cut is sealed quickly**?**

1 Copy and complete:
The liquid part of the blood is called the . . . .
The red cells contain a substance called . . . .
This substance helps the . . . . cells to carry . . . . The white cells . . . . the body from germs. One kind of white cell . . . . germs.
Another type makes . . . . that kill germs.

2 Make a table of the differences between red and white blood cells.

3 People who live at high altitude have far more red cells than you. Why do you think this is?

4 Besides our blood, what other protection do our bodies offer against infections?

Things to do

Self-defence

How is it that our bodies are able to fight off disease? If you catch a disease like measles, you don't get it again – you become **immune**.

Why do you think this happens? Write down your ideas.

Fighting off the enemy

Do you remember how white blood cells defend your body?

All germs (that's bacteria and viruses) have chemicals on their surfaces. These chemicals are called **antigens**.
When you catch a disease like measles, the white blood cells in your body make chemicals called **antibodies**.
These antibodies attach themselves to the antigens on the surface of the germs.
They cause the germs to clump together and make them harmless.
Now another type of white blood cell, let's call them 'killer cells', are able to 'eat' the 'stuck-together' germs.

After any disease, the antibodies stay in your blood, making you immune.

a So how can you become immune to some diseases and not others?

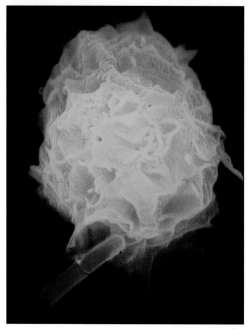

A white blood cell attacking an *E.coli* bacterium

Each germ has antigens of a particular shape. So you produce antibodies that match the shape of each antigen. This is why you have to catch the disease before you become immune to it.

1. Bacteria enter your body through a cut in your skin.

2. Your body makes antibodies to fight the invaders.

3. The bacteria are destroyed by the antibodies.

4. Antibodies stand by ready to fight off any future attack. The body is **immune**.

▶ Study the cartoons and use the information to write a brief explanation of what happens when harmful bacteria invade your body.

You can also become immune to a disease by **vaccination**. A vaccine is a weak form of the disease microbe.

b Explain how you think vaccination works.
c Your body can make millions of **different** antibodies. Why do you think this is?
You can find out more about vaccination on pages 40 and 41.

d Find out how **antiseptics** kill germs outside your body.

Disinfectants are strong chemicals that kill germs on floors and work surfaces.

e Why are **disinfectants** not used to kill germs on or in our bodies?

A mother passes on antibodies to her baby through breast milk. Why is this important?

Mr Clean

Bob is the caretaker at the high school.
He's got a problem.
The label has come off his big bottle of disinfectant.
So he doesn't know how much to add to water before use.
If he adds too little, it won't kill the germs.
If he adds too much, it'll be expensive.
Help Bob out by finding the **smallest** amount of disinfectant that will kill the germs.

(Hint: You could use agar plates and paper discs with different concentrations of disinfectant on them.)

Your teacher will give you an agar plate with harmless bacteria growing on it.

How much disinfectant will you add?

How will you make it a **fair and safe test**?

How long will you leave it to work?

How will you record your results?

Ask your teacher to check your plan, then try it out.

do not open
plate after incubation

---

1 Copy and complete:
If you catch a disease and you don't get it again, you are . . . . to it.
Your body makes chemicals called . . . .
They stick the germs together and make them . . . .
The . . . . stay in your blood to give you immunity to the disease. A . . . . is a weak form of the disease microbe. It can be . . . . into your body or taken by mouth. It gives you . . . . to a disease.

2 Find out whether the following are true or false.
a) Tetanus is caused by germs getting into an open wound.
b) A pregnant woman cannot pass on rubella to her unborn child.
c) Smallpox vaccine is no longer given because the disease has been wiped out.
d) There is a low risk of whooping cough vaccine causing brain damage to some babies.

3 Why is it important that the following places are free from germs?
a) Swimming pools.
b) School kitchens.
c) Doctor's surgeries.

4 Try to find out what diseases people can be vaccinated against.
Which vaccines have you been given?

A rubella vaccination

Things to do

The spread of disease

Learn about:
● early ideas about disease
● how our ideas change

1600

Can you remember what microbes are?

400 years ago nobody had heard of microbes.

They didn't even know where other living things came from.

"Look, this meat is full of maggots!"

"Yes, the meat has created the maggots."

We now know that food rots because of microbes.
In the old days, most people thought rotting food *made* the microbes.

"The mutton gravy has changed into new life."

John Needham believed in this **spontaneous creation**. He thought that when an animal died parts of it formed new creatures.

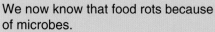

BOILED

Lazzaro Spallanzani showed that food does not go bad if the microbes are killed. He killed the microbes by sealing the food and then boiling it.

In 1854 **Louis Pasteur** isolated microbes and added them to sterilised soup – they multiplied.

It's off!

He was able to show that it was microbes that made wine and milk decay.

There is a vital **force** in all living things. When they die it produces microbes.

Felix Pouchet still believed in spontaneous creation. He came to opposite conclusions to Louis Pasteur.

I boiled the broth in each flask.

The microbes can't get into this flask. They get stuck in the neck.

Pasteur's experiments were more reliable than Pouchet's. He proved that microbes caused diseases such as anthrax.

In the nineteenth century, hospitals were not very clean places. Microbes spread easily and wounds often became infected.

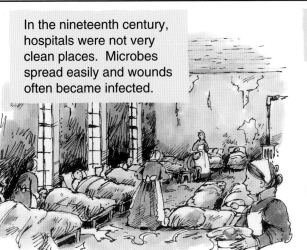

Women giving birth in hospitals sometimes died of fever afterwards. **Dr Ignaz Semmelweiss** noticed that doctors never washed their hands between patients.

Our doctors are carrying the infection on their hands.

They must wash their hands in disinfectant between patients.

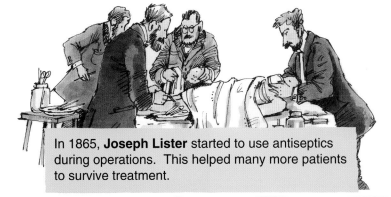

In 1865, **Joseph Lister** started to use antiseptics during operations. This helped many more patients to survive treatment.

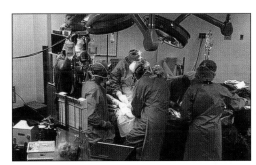

Modern operating theatres are kept as free from microbes as possible.

1 Write down what you can remember about the 3 main types of microbes.

2 What happens if a bottle of milk is left open to the air?
Give your answer as if you agreed with John Needham and Felix Pouchet.

3 a) How was Louis Pasteur able to kill microbes in milk?
b) What do we call this process today?
c) Name 4 other ways in which microbes in food can be killed.

4 a) How did Semmelweiss discover that disease was spread in hospitals?
b) How was he able to reduce this spread of disease?

5 Look at the operating theatre in Lister's time.
a) What conditions can you see that reduce the spread of infection?
b) What things do not reduce this spread?
Now look at the modern-day operating theatre.
c) What precautions have been made in it to prevent the spread of infection?

Things to do

Microbes and health

Ideas about vaccination

In the eighteenth century, a farmer called **Benjamin Jesty** noticed that people who had caught cowpox did not get smallpox. Smallpox is a far more dangerous disease than cowpox. He scratched the skin of each member of his family and put some cowpox material on the wound. This prevented them from getting smallpox when there was a major outbreak of the disease in 1774.

In 1776, **Edward Jenner**, a physician in Gloucestershire, carried out a medical experiment. It is reported in this newspaper article:

Jenner vaccinating his son

THE GLOUCESTER TIMES

BOY SURVIVES SMALLPOX

1776

DOCTOR EDWARD JENNER has carried out a reckless experiment on a young boy.

He scratched some liquid from a smallpox blister into the arm of 12-year-old James Phipps. Smallpox is responsible for many deaths every year. How is it that the parents gave their consent? It seems to be a miracle that the boy has survived.

Doctor Jenner puts it down to "Scientific observation"! Apparently he has noticed that milkmaids never catch smallpox.

Although they often catch cowpox, a mild disease. He told our reporter, "I took some pus from a cowpox blister and scratched it into the arm of young James. He later developed cowpox, but soon recovered. Later I inoculated him with smallpox, but he did not show signs of the disease – I believe that he is now immune to it."

Where will all this experimenting end?

Will we all start growing horns as a result of the new cowpox inoculation?!!

▶ Use reference material and the internet to find out about routine immunisation programmes. Write a magazine article about the advantages and disadvantages of this strategy.

Acquired immune deficiency syndrome (AIDS)

AIDS is caused by HIV virus which attacks the body's immune system. HIV attacks the white blood cells that protect us.

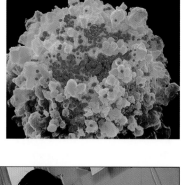

HIV virus attacking a white blood cell

a   How could HIV be detected in a person?

HIV is only transmitted if the blood or semen (fluid which contains the sperms) of an infected person enters the blood stream of another person.

b   How might this occur by the following:

i)   the transfusion of infected blood?

ii)   drug addicts sharing needles?

iii)   sexual activity with many partners?

iv)   In each case above, suggest how the spread of HIV can be prevented.

There is no cure for AIDS and as yet no vaccine for HIV. AIDS has spread very quickly in parts of Africa.

c   Why do you think this is?

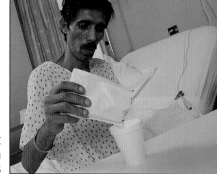

A patient suffering from AIDS

Infectious disease

d Name 3 groups of microbes that can cause infectious disease.

e Give 3 ways in which these microbes can be spread from one person to another.

The early stage of a disease is called the **incubation period**.

f What is happening to the disease microbe inside the body during this time**?**

Disease microbes can do damage to tissues and release **toxins** which cause the **symptoms** of a disease.

g Give some examples of common disease symptoms.

Look at the graph:

h What symptom of the disease is shown on the graph**?**

i How long was the incubation period**?**

j What caused the fever and how long did it last**?**

Antibiotics are used to relieve disease symptoms, but many microbes are becoming resistant to their action.

k How do you think a microbe can develop resistance to a particular antibiotic**?**

In Scotland, in 1997, a new strain of *E.coli* 0157 caused an outbreak of food poisoning that affected 500 people and killed 20.

Look at the newspaper report:

l How do you think the disease was spread**?**

m Why did the existing antibiotics prove useless**?**

n Find out about other diseases which are increasing due to increased resistance to antibiotics.

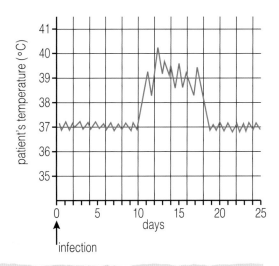

Death toll rises in food bug outbreak

Five people have died of food poisoning in Britain's worst case of *E-coli bacteria* contamination.

A hospital has been closed to all GP-arranged admissions except suspected cases of the *E.coli 0157* food poisoning outbreak.

The butcher's shop thought to be the source of the outbreak announced yesterday that is was temporarily closing.

Seven members of staff linked to the food poisoning outbreak in Scotland are infected.

Thirty-two adults and a child were being treated yesterday in the hospital, where the Lanarkshire Infectious Disease Unit is based. The number giving cause for concern rose from ten to 15 over the weekend, and the number showing symptoms rose from 189 to 209.

Things to do

1 a) How do you think cowpox is different from smallpox?

b) How do you think that cowpox is caught?

c) What do you think gave Jenner the idea of inoculating against smallpox?

d) Explain why you think James did not catch smallpox.

e) Try to explain the reporter's attitude to Jenner's experiment.

2 In the nineteenth century, **Louis Pasteur** showed that he could heat anthrax bacteria and make it safe. When he then injected the vaccine into sheep, it gave them immunity to anthrax.

a) How do you think that heating the anthrax bacteria made them safe?

b) Why do you think that injecting the 'safe' anthrax bacteria gave the sheep immunity? (Hint: try to use the words 'antigens' and 'antibodies' in your answer.)

3 Poliomyelitis ('polio' for short) is a virus that destroys nerve cells. It can damage the spinal cord causing the victim to become paralysed.

In 1953, **Jonas Salk** made a vaccine that prevented polio. It was so successful that polio has now disappeared from developed countries.

Use books, ROMs and the internet to find out more about the work of Jonas Salk and how he developed the anti-polio vaccine. Include any data you find on the numbers contracting the disease before and after the work of Salk.

Mothers queue up to have their children vaccinated against polio in the mid-1950s

Questions

1 Decomposers, like fungi, are very useful in the environment.
Explain why plants and animals depend upon them so much.

2 The graph shows the growth of one bacterium to
64 million over 15 hours:
  a) What is happening to the bacteria over the first
     3 hours?
  b) What is happening to the bacteria between 4 and
     12 hours?
  c) When did the growth of the bacteria begin to
     slow down?
  d) What could have caused the growth of the bacteria
     to slow down?
  e) How do you think the numbers of bacteria were
     estimated every hour in this experiment?

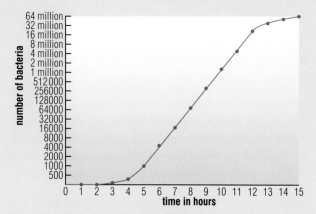

3 Milk can be preserved in a number of ways:
  a) pasteurised   b) sterilised   c) ultra-high temperature (UHT).
Find out about each of these methods.
Why does pasteurised milk turn sour even in a sealed bottle?
Why do sterilised and UHT milk keep so long in their containers?

4 George says: "when I dig my compost, steam comes out of it".
Plan an investigation to find out how much heat is given off by
rotting grass.
You can use the sort of equipment found in your science laboratory.
Remember to make your investigation a fair test. Do not try out your
plan unless you have checked it with your teacher.

5 Look at the picture. It shows some of the antibodies that Martin
has in his blood.
  a) Does Martin's blood contain antibodies to fight polio?
  b) Is he immune to polio?
  c) Can he catch polio?
  d) Does Martin have the antibodies in his blood to fight measles?
  e) Is he immune to measles?

antibodies

polio
virus

measles
virus

6 Equal amounts of two different bacteria were grown in the same
conditions after being placed on agar containing different
concentrations of antibiotic. The results were recorded after
the bacteria were kept in an oven at 25 °C for 6 days.
  a) Plot the results as line-graphs.
  b) What was the colony diameter of A in antibiotic
     concentration 1.75%?
  c) What was the minimum concentration needed to kill B
     completely?
  d) What characteristic of living things does the colony
     diameter measure?

% concentration of antibiotic	Colony diameter (mm)	
	Bacteria A	Bacteria B
0.1	63	60
0.5	55	45
1.0	45	30
1.5	35	17
2.0	25	5
2.5	15	0
3.0	0	0

Ecology   8D

Living things depend upon their environment to survive.

They have to compete for things like food and space.

If they are well adapted then they have more chance of surviving and their numbers will increase.

In this unit:

Classifying

Learn about:
● classifying living things
● the 4 main plant groups

Animals

Here is a reminder of the way that scientists classify the animal kingdom:

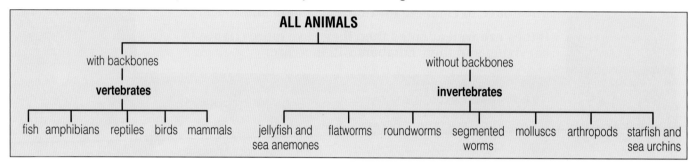

ALL ANIMALS

with backbones — **vertebrates**: fish, amphibians, reptiles, birds, mammals

without backbones — **invertebrates**: jellyfish and sea anemones, flatworms, roundworms, segmented worms, molluscs, arthropods, starfish and sea urchins

a   What are the 4 sub-groups that arthropods can be divided into? (See Book 7, page 56.)

Plants

What are plants? You already know that they are very different from animals. For one thing they make their own food.

For this reason, plants are often known as **producers** of food, whilst animals are **consumers** of food.

As with animals, you can put plants into groups to help you find their names.

ALL PLANTS

These don't have seeds:

Mosses
Weak roots
Thin, delicate
leaves

Ferns
Strong stems,
roots and
leaves

These have seeds:

Conifers
No flowers
Seeds produced
in cones

Flowering plants
Have flowers
Seeds produced
inside fruits

Mosses

● Live in damp places.
● Have thin leaves that easily lose water.
● Make tiny **spores** instead of seeds. These are carried away by the wind. Moss spores grow into new moss plants.

▶ Look at the picture or some moss plants:
b   Where do you think the spores are made?
c   How heavy do you think the spores will be? Give your reasons.
d   Why do you think that mosses are only found in damp places?

Ferns

● Have strong stems, roots and leaves.
● Make spores instead of seeds.
● Have tubes that carry water around inside the plant. The tubes are called **xylem**.

▶ Look at a fern leaf:
e   Where do you think the spores are made?
f   How are they protected from the rain?

Conifers

- Many are evergreen with leaves like needles.
- Have xylem tubes.
- Their seeds are produced inside **cones**.

▶ Look at a pine cone. Can you find the seeds inside it?

g How do you think these seeds are carried away?

Flowering plants

- Produce flowers.
- Have xylem tubes.
- Make **seeds** inside fruits and berries.

▶ Try cutting open a broad bean seed or maize seed. Look for a **very young plant** and its **food store** surrounded by a **hard seed coat**.

h Write down your ideas about what each of these parts do.

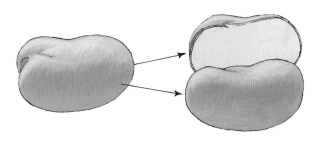

It's a fact!

The largest living plants are the giant redwoods of North America that grow over 100 metres high.

---

1 Copy out and complete the following table.

Group	Do they have strong stems, roots and leaves?	Seeds or spores?	Flowers or not?
mosses ferns conifers flowering plants			

Things to do

2 Write down some examples of how plants have become important as each of the following:
a) food
b) fuel
c) medicine
d) building materials.

3 Over three million years ago, in what was known as the Carboniferous age, vast forests covered the Earth. The plants were similar to the ones in this photograph. They had stems, roots and leaves. These plants reproduced by means of spores. Which group of plants do they belong to?

Competition

Learn about:
- the importance of sampling
- using quadrats
- competition for resources

What does the word **competition** mean to you?

A race is a competition. Everyone tries very hard to win.
But there can only be one winner.

In nature, living things compete for **resources** that are in short supply e.g. food and space.
Those that compete successfully will survive to breed.

▶ Write down some of the resources that animals compete for.

Write down some of the resources that plants compete for.

Seeing red

Robins compete for a **territory** (habitat) all the year round. They sing to let other robins know that the territory is occupied. During the breeding season they build a nest and raise their young inside the territory. At this time they are very fierce and drive other robins away.

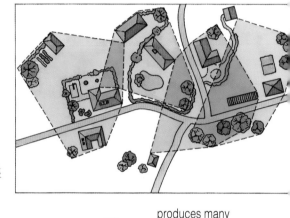

a How many robin territories are shown on this map?

b What resources are the robins competing for?

c Why do you think that robins are so fierce to other robins but not to all birds that enter their territory?

Weed this!

A weed is a plant that is growing where it is not wanted e.g. poppies in a wheat field.

It's easy to see why gardeners and farmers hate weeds and try hard to get rid of them.

Weeds compete with the other plants for light, water and space.

The dandelion is a successful weed.
Can you see why?

▶ Look at the adaptations of the dandelion in the diagram.

Copy and complete the table:

grows quickly and flowers twice a year

produces many seeds which are spread by the wind

resistant to many weedkillers

seeds germinate rapidly

leaves spread out over ground

grows quickly on bare soil

roots produce chemicals that stop other plants growing

deep roots which are difficult to remove

Adaptation	How it helps the dandelion to survive
Seeds germinate rapidly	Quickly makes new weed plants

Competition on the playing field

Dandelions, daisies and plantains compete with grass on your school field.

How could you find out which is the most successful weed?
You could count all of them, but this would take a very long time!
Instead you could take a **sample**. You could count the numbers of each weed in a small square called a **quadrat**.

1 Put your quadrat down on a typical area of the school field.
2 Count the numbers of dandelions, plantains and daisies inside your quadrat.
3 Take 4 more samples in different parts of the school field.
4 Record your results in a table like this:
* Add the totals of the class together for each weed.
* Draw a bar-chart of the class results.

Weed	Sample					
	1	2	3	4	5	Total
dandelions	3	3	4	0	2	12
plantains						
daisies						

d Why were you asked to take 5 samples?
e Which weed was the most successful on your school field?
f Try to think of a hypothesis that could explain this.
g What further investigation could you do to test this hypothesis?

1 Copy and complete:
Living things . . . . for resources that are in . . . . supply, such as . . . . and . . . . Those plants and animals that . . . . successfully will . . . . to breed. Weeds compete with crops for . . . . and . . . .

2 Here are the planting instructions on a packet of broad bean seeds:
Sow the seeds about 5 cm deep and about 20 cm apart in open ground.
a) Why shouldn't the seeds be planted:
   i) any closer together? and
   ii) any further apart?
b) What resources might these plants compete for?

3 Can you think of any animals or plants that compete with humans?
Many of those that compete with us for food we call **pests**.
Write down any that you can think of and say what you think they compete with us for.

locust

Different habitats

Learn about:
● the variety of habitats
● comparing different habitats

Do you remember what a **habitat** is?
A habitat is a place where animals and plants live.

▶ Write down some habitats and the animals and plants that live in them.

Woodland and grassland

Think about what conditions are like inside a small wood.

a What will the light, humidity and temperature be like?

b How will these conditions change during the year?

c Apart from the trees, what other plants will grow here?

d What sort of animals would you find and where would they live inside the wood?

e Can you think of any adaptations that the woodland plants and animals might have in order to survive?

Think about what conditions are like in rough grassland.

f How will the environmental conditions differ in grassland?

g What sort of animals will you find living in grassland?

h Apart from the grasses, what other plants would you find?

i Can you think of any adaptations that grassland plants and animals might have to help them to survive?

Rivers and ponds

Think about what conditions are like in a river.

j What will the rate of flow of the water be like?

k How will this affect the amount of oxygen dissolved in the water?

l What sort of adaptations will small invertebrate animals need to help them to survive in a river?

Think about what conditions are like in a pond.

m What is the flow rate of the water like?

n How will the light intensity and temperature change with the depth in a pond?

o What will the plants be like in a pond and where will you find them growing?

p Make a list of some common pond animals that you might find living among the plants.

caterpillar feeding on willow

great diving beetle

Comparing habitats

Carry out a study of the communities in *two* local habitats.

Depending upon your school location, you might choose to study:

- 2 different areas of grassland on a playing field
- a hedge, and a ditch
- a path, and an area of untrodden grassland
- a pond, and a stream
- 2 different areas in a woodland.

Always work supervised by your teacher, and take particular care when working near water.

For your study:

- Collect data relevant to your habitats using ICT, for example, temperature, light intensity, dissolved oxygen. You may want to measure pH, temperature, moisture, organic content and mineral content if you are studying soils.

- Sample some of the plants and animals in the *two* habitats and suggest how they are adapted to survive.

- Write a report, including graphs of your data, comparing conditions in the *two* habitats and the adaptations of the plant and animal communities.
  Evaluate your method and explain how you arrived at your conclusions.

You may want to make a presentation of your report to the rest of the class or present your work as a wall display.

Sampling plants found across a path

Measuring the pH of a pond

Rocky shore at low tide

Things to do

1 Copy and complete:
A . . . . is a place where plants and animals live. Habitats vary due to different . . . . factors such as temperature, ... intensity and dissolved . . . . Those . . . . and plants that are best . . . . to their environment will . . . . and produce offspring.

2 In any environment there are factors that affect the survival of plants and animals. Give an example of how each of the following factors can limit the growth of a plant or animal population:
a) predation
b) disease
c) competition for food
d) competition for light.

3 Search for sources of secondary data about plants and animals found in *two* different habitats, e.g. rocky and sandy shores, different stages in sand dunes, lakes and streams, different urban habitats.

4 For the habitats you have studied, make a table of the different organisms that you found and show how they were adapted to survive in each habitat.

Population growth

A **population** is a group of the same animals or plants living in the same habitat e.g. greenfly on a rose bush or daisies in a lawn or a shoal of herring in the sea.

a Write down some more animal and plant populations and the habitats that they live in.

b Why do you think that animals and plants live together in populations?
Think about what they need for survival and how they keep up their numbers.

How do populations grow?

What happens if rabbits colonise a new area?
First there are a few of them and plenty of food.
Many will survive to breed and their numbers increase.
At first there are no predators to keep their numbers down.

Soon each generation is double the size of the previous one!
8 becomes 16, 32, 64, 128, 256, 512, and so on.
But this can only take place under ideal environmental conditions.

So why isn't the Earth over-run with rabbits?
The rabbit population can not go on increasing for ever,
because not all of them will survive.
Some factors will start to *limit* the population growth, such as lack of food or being eaten by predators.

Here are some *limiting factors* that can slow down growth of populations:

● light   ● overcrowding   ● food and water   ● disease
● climate   ● predators   ● oxygen   ● shelter

▶ Copy and fill in the table for each factor above.

Factor	How the factor can limit population size
Light	lack of light slows growth of plants

People often think of populations as only being plants or animals.
But you can have populations of microbes too.
If you are ill you may have a population of bacteria or viruses inside your body!

The photo shows part of a yeast population:
For thousands of years, yeast has been used to change the sugar in barley into beer and the sugar in grapes into wine.
It also reacts with the sugar in dough to produce carbon dioxide.
This makes the dough rise to give us bread.

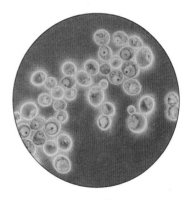

yeast cells

The growth of a yeast population

You can grow yeast cells in a sugar solution.

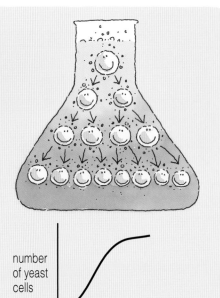

Add a suspension of yeast cells to the sugar solution in a conical flask.
Swirl the flask and plug its opening with cotton wool.
Place the flask in a warm cupboard or incubator at 20–25°C.

Over the next few hours look at a drop of the yeast under a microscope.
Make sure that you use the same magnification each time.
Count the number of yeast cells that you can see and write it down.

After a while you may be able to draw a graph like the one here:

c What do you think happened to the number of yeast cells
   i) in the first hour?   ii) later in the experiment? and
   iii) at the end of the experiment?

d Why do you think the population of yeast cells stopped growing?

e What factors could you change in this experiment to increase or to decrease the growth rate of the yeast population?

Human effects

Humans can limit the growth of populations.
In many parts of the world large areas of forest are being cut down.
The trees are used for timber or the land is cleared for farming.

Trees should only be cut down if they can be replaced naturally or by planting. This would result in **sustainable** growth.

In a similar way fishermen are taking too many fish out of the sea.
Modern technology, such as the use of sonar to find the fish, more powerful ships and huge plastic nets, means more fish are caught.
Many species are **overfished** and in danger of extinction.

We should limit fishing so that enough fish survive and reproduce.

f What could be done to control the amount of fishing?

---

1 Copy and complete:
A . . . . is a group of animals or plants living in the same . . . . The growth of a . . . . can be limited by factors such as . . . . , ... and . . . . , so not all the young animals or plants in a population will . . . . to breed.

2 One of the slowest-breeding animals is the elephant. It has been worked out that, starting with one pair of elephants, their offspring would number 19 million after 700 years.
Explain why this could never actually happen.

3 Explain how you think each of the following would affect human population growth:
a) famine and disease
b) high birth rate
c) improved medical care.

Things to do

Exploring pyramids

Learn about:
- pyramids of numbers
- factors affecting populations
- energy flow in a food chain

Food chains can show how food (and energy) pass from one living thing to another.

Write out this food chain in the correct order:
owl, oak leaves, shrew, caterpillar

a Where do the oak leaves get their energy from**?**

How many?

Food chains cannot tell you **how many** living things are involved.
It takes lots of leaves to feed a caterpillar, and lots of caterpillars to feed one shrew.

▶ Look at the diagram:

 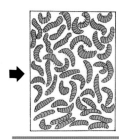 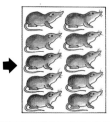 

flow of energy

b Why are there more plants than there are herbivores**?**

c Why are there more prey than there are predators**?**

Up the pyramids

Look at the numbers in this food chain:

owl	1
shrews	10
caterpillars	100
oak leaves	300

You can show this information in a **pyramid of numbers**.
The area of each box tells us how big the numbers are.
Start with the plants on the first level and build it up:

level 4

level 3

level 2

level 1

flow of energy

▶ Copy the diagram and label each feeding level.

d What happens to the **numbers** of living things as you go up this pyramid**?**

e What happens to the **size** of each living thing as you go up this pyramid**?**

f Why are the plants always on the first level**?**

▶ Now try drawing a pyramid of numbers for each of these food chains:

	producer	*herbivore*	*carnivore*	*top carnivore*
g	5 cabbages	20 slugs	5 thrushes	1 cat
h	1 oak tree	100 caterpillars	5 robins	100 fleas

A top carnivore

52

Funnel fun

You can find very small animals in leaf litter using a **Tullgren funnel**.

- Set up the funnel as shown:
- Place a sample of leaf litter onto the gauze.
- Switch on the light and leave it for 30 minutes to work.
- *Be careful not to over-heat your sample and kill your animals.*
- Use a lens to look at the tiny animals that you have collected.

i What two things made the animals move downwards?

j Why should the light not be too close to your leaf litter?

k Why should your layer of leaf litter not be too thick?

Your teacher can give you a Help Sheet to identify your animals. It also tells you what they eat.

- Make a count of each of your animals and record it in a table:

Phew!... too hot and dry up there for me.

Animal	Number found	Herbivore or carnivore?
mites springtails symphylids		

- Draw a pyramid of numbers for the animals you have collected.

---

1 Copy and complete:
Pyramids of . . . . can tell us about the numbers of living things at each feeding . . . .
Plants are always put in the . . . . level because they make food by . . . . and so bring energy into the food chain.
As you go up the pyramid, the numbers of animals get . . . . and the size of each animal gets . . . .

2 a) Draw a pyramid of numbers from these data:

       sparrowhawk   1
       blue tits   5
       bark beetles   50
       beech tree   1

b) How is it that one tree can support so many herbivores?
c) Why do you think this is called an 'inverted pyramid'?

3 Instead of using numbers to draw pyramids, scientists sometimes use **biomass**. This is the weight of living material.
a) What would the pyramid of numbers in question 2 look like as a pyramid of biomass?
b) Draw and label it.

4 Look at the Help Sheet that you used in the activity on this page.
Choose 5 of the animals and make a key that you could use to identify each one.
Get a friend to try it out.

Things to do

Poison algae

Learn about:
● factors affecting populations
● living pollution indicators

Algae are tiny plants. They live in lakes, rivers and seas.

Look at this food chain:

algae ⇒ water fleas ⇒ small fish ⇒ large fish ⇒ people

a Why are algae found at the start of the food chain**?**

b Can you name 3 things that algae need to grow**?**

c Why are fisheries often found where there are lots of algae**?**

Algae grow well when there is light, warmth and lots of **nutrients**.
Nutrients, like **nitrates** and **phosphates**, make algae grow best.

d What do you think will happen if *lots* of nutrients get into the water**?**

▶ Look at the picture:
Write down the ways in which *extra* nutrients get into the river.

Too much of a good thing

More nutrients mean more plant growth – in this case, lots of algae.
This can cause problems for animals and other plants in the water.

▶ Write out these sentences in the correct order.
They will tell you what can happen when too much algae grows.

Algae die and sink to the bottom of the lake or river.	Extra nutrients in the water make the algae grow fast.	The lack of oxygen in the water kills fish and other water animals.	Dead algae are broken down by microbes which use up lots of oxygen.

A soapy story

Detergents contain phosphates. These are plant nutrients.
In this experiment, you can find out the effect of detergent on the growth of algae.

● Label 6 test-tubes 1 to 6.
● Add 5 cm³ of nutrient solution to each tube.
● Using a clean dropper, add 5 drops of algae water to each tube.
● Using clean droppers, add the following amounts of detergent solution and distilled water.

Test-tube	Drops of detergent solution	Drops of distilled water
1	0	5
2	1	4
3	2	3
4	3	2
5	4	1
6	5	0

● Leave all 6 test-tubes in a well-lit place.
● After a few days, shake each tube and see how green it is.
Compare the growth of algae in each one.
● Record your results in a table.
● Discuss your results and write down your conclusions.

Living indicators

We can use water animals to tell us how pure the water is:

- Come on in, the water's fine!

Mayfly larvae and stonefly larvae need clean water.

- It's getting worse!

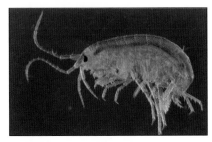 

Freshwater shrimps and water-lice can stand some pollution.

- The dirty duo!

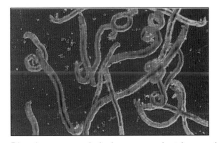 

Blood worms and sludge worms just love pollution!

e If you find a stonefly larva in a water sample, what would it tell you about the water?

f Which animals do you think could survive in the water in test-tube 6 in your investigation?

g Can you think of any other ways of testing the water for pollution?

---

1 Copy and complete:
Algae are . . . . that live in water. To grow well, they need light, warmth and . . . .
Two of the nutrients which increase the growth of algae are . . . . and phosphates.
If too much algae grow, they die and . . . . rot them down. The microbes use up a lot of . . . . and this means that fish and other water animals will . . . .

2 Are farmers poisoning our water supply?
Some of the nitrates in fertilisers wash out of the soil and trickle down into the bed-rock. Very slowly, the nitrates are moving nearer to water that we use to drink.
a) Try to explain the 'nitrate time bomb'.
b) Find out what effects nitrates can have on our health.

3 a) Should chemical fertilisers be banned?
b) What would happen to the world's food production if they were?
c) What could we use instead of them if they were banned?

Clearing weeds from a canal

Things to do

55

Ecology

Learn about:
● the development of ideas
● how scientists worked in the past

Charles Elton was responsible for many of the ideas that formed the science that we call **ecology**.

Ecology is the study of how living things interact with each other and with their environment.

In 1921, he took part in an expedition to Spitsbergen in the Arctic Circle and made a survey of the animal life there.

As a result, in 1927 he wrote his classic book *Animal Ecology* in just 3 months!

In it he set out his ideas about food chains, food webs and pyramids of numbers.

These ideas were based upon his close observations of the animals that he had seen in the Arctic.

But Charles also introduced scientific method into his study of natural history.

His conclusions were not based upon his observations alone; he also made counts of the numbers of animals found in an area.

Some members of the expedition to Spitsbergen

A young Charles Elton

Food chains

Elton first pieced together **food chains** based upon the feeding relationships that he observed.
Here is an example:

plant plankton ➡ krill ➡ squid ➡ leopard seal ➡ polar bear

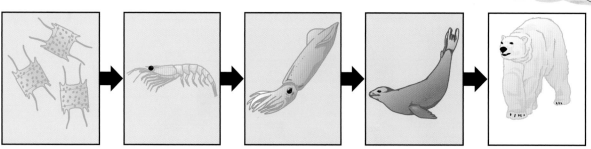

The plant plankton is eaten by the krill, which in turn is eaten by the squid, which is eaten by the seal, which is eaten by the polar bear.

Elton noticed that predators are usually larger than their prey. So organisms often increase in size as you go along the food chain.

Food webs

But communities are not made up of just one food chain.
For instance, krill also form the food of penguins, fish, seals and a number of whales.
So a more realistic picture of the feeding relationships in a community can be shown in a **food web**.

Elton was able to piece together a complex food web for the living organisms on Bear Island.
A food web shows a number of interlocking food chains.
Here is an example of a seashore food web:

Food webs do not tell us **how many** individuals are involved in a community.
Elton noted that, 'animals at the base of the food chain are relatively abundant, while those at the end are relatively few, and there is a progressive decrease between the two extremes'.

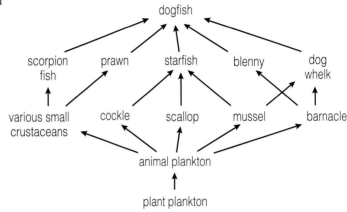

Pyramids of numbers

Elton used the number of individuals in a community to construct **pyramids of numbers**.

Unlike food chains and webs, pyramids of numbers show the actual number of organisms involved at each feeding level.

Elton took samples of the organisms in a community and divided them into producers, herbivores and carnivores. He drew a horizontal bar-chart, with the area of each bar being proportional to the number of individuals at each feeding level.

Things to do

1 What do you think is meant by the word 'ecology'?

2 a) What are food chains able to show?
b) In the arctic food chain, name:
 i) the producers
 ii) the herbivores
 iii) the top carnivores.

3 Why do food webs give 'a more realistic picture of the feeding relationships in a community' than do food chains?

4 a) What is meant by a pyramid of numbers?
b) What advantages do pyramids of numbers have over food chains?

5 Look at the seashore food web above.
a) What are the producers?
b) What do starfish feed on?
c) If the dog-whelks were all killed by pollution, what would happen to the populations of
 i) mussels? and
 ii) animal plankton?

6 Try to find out more about the life and work of Charles Elton using books, ROMs or the internet.

Questions

1 A large herd of deer lived on an island.
The deer were sometimes killed by predators such as wolves.
To protect the deer population, some hunters shot all the wolves.
Then the deer population grew. Their numbers became so large
that they began to compete for grass. Many deer starved.
Soon the deer population was about the same as it was before
the wolves were shot.
a) What were the 4 populations involved?
b) Why do you think the hunters were wrong to shoot all
the wolves?
c) What do you think will happen to the deer population in
the future?

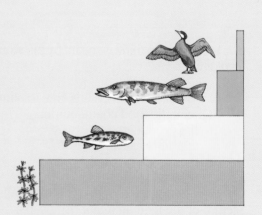

2 "Big, fierce animals are rare." Try to explain this statement.

3 Look at the way the leaves of these weeds are growing.
a) How do you think they survive trampling?
b) How do you think they affect the growth of the grass around
them?
c) In what ways do gardeners cut down the competition from
these weeds?

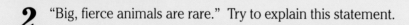

plantain   daisy   dandelion

4 The following data were collected from a river:

pike	1
trout	10
water fleas	500
algae	10 000

a) Draw a pyramid of numbers (not to scale) for the river.
b) Which living thing would you remove if you wanted to increase
the number of trout in the river?
c) Give one other effect of removing this living thing from the river.

5 Some DDT was sprayed on a lake to control mosquitoes.
Look at the table showing the amounts of DDT in a food chain.

cormorant	26.5 ppm
pike	1.3 ppm
minnow	0.2 ppm
algae	0.05 ppm
water	0.000 05 ppm

Explain how the cormorant has 500 000 times more DDT in its
body than there is in the water.

6 The Royal Society for the Protection of Birds has said that more than
40 bird species are threatened by intensive farming. Birds in decline on
farmland include the skylark, barn owl, lapwing and golden eagle.
a) List the ways in which you think farming can reduce bird numbers.
b) What steps do you think could be taken to improve this situation?

Atoms and elements

Your life is full of elements.
You are made of them.
You eat them. You drink them.
You are surrounded by them.

In this unit you can find out more about elements.
What are they made of?
What can we make from them?

Chemical elements

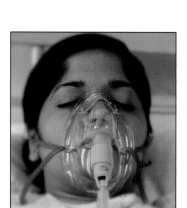

▶ Write down your ideas about the following questions:

a How do you know that the air is all around you?

b What is passive smoking? Why do people worry about this?

c Something is cooking in the kitchen!
Why can you smell it at the door?

In Book 7 you found out that everything is made from particles.
The particles are invisible. They are very small.
They are called **atoms**.
Atoms are the smallest parts of any substance.
They make up solids, liquids and gases.
If a substance is made of only one type of atom, it is a *simple substance*.
We call it an **element**. **Elements** can be solids, liquids or gases.
Look at the examples below:

A copper coil

Mercury liquid

Oxygen gas

Elements are substances which *cannot be broken down into anything simpler*. **Elements** have *only one type of atom*.

You can show atoms like this:

oxygen atoms   All the atoms which make up oxygen gas are the same as each other.
Oxygen is an **element**.

nitrogen atoms   All the atoms which make up nitrogen gas are the same as each other.
Nitrogen is an **element**.

. . . *But* remember . . . oxygen atoms are different to nitrogen atoms.
Oxygen and nitrogen are *different* elements.

Your body is made up of many elements. Some of these are:

calcium	carbon	chlorine	hydrogen
magnesium	nitrogen	oxygen	phosphorus
potassium	sodium	sulphur	

Mostly these elements are not found on their own.
They are found in **compounds** in your body.
Compounds are substances which have 2 or more elements joined together. They have 2 or more different types of atom.

Compounds look different to the elements they are made from.

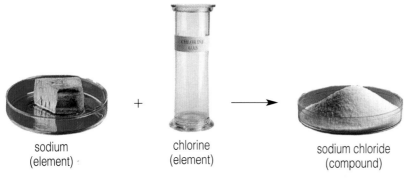
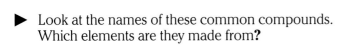

sodium (element) + chlorine (element) → sodium chloride (compound)

Adding sodium to water

Compounds can also **behave** differently to the elements they are made from.

eye protection and safety screen

▶ Watch carefully as your teacher adds the **elements** sodium and chlorine to water. What do you see?
What happens when sodium chloride, a **compound**, is added to water?

▶ Look at the names of these common compounds.
Which elements are they made from?

sodium chloride   carbon dioxide   hydrogen oxide (water)

How many elements?

d Make a list of all the elements mentioned so far in 8E1.
How many elements are there in your list?

e Use a reference book to find a copy of the periodic table.
This lists the chemical elements in a special order.
How many elements are shown in the periodic table?

f Explain why we can have millions of different compounds made from a relatively small number of elements. The Lego pictures opposite may give you a clue!

1 Copy and complete:
All substances are made of very small particles called . . . .
Substances which contain only one type of . . . . are called . . . .
. . . . cannot be broken down into anything simpler. When these combine together, they make . . . .

2 Look at labels on foods at home. Make a list of elements and compounds that the foods contain.

3 If you discovered a new element, what would you call it?

4 Say whether each of the following is an element or a compound:

chlorine   magnesium   iron
sulphur dioxide   sulphur
carbon dioxide   iron chloride
calcium   carbon   sodium

5 The first scientist to suggest the name element was Robert Boyle. The year was 1661. Find out some information about Boyle.

Things to do

Simple symbols

Often a symbol or picture gives information quicker than lots of writing.

▶ Test yourself on the following examples.
What do the symbols mean?

Elements

Chemists have a shorthand way of writing about elements.
They use **symbols** instead of writing out the names.

▶ Copy out the table. Fill in your guess for each element's symbol.
Then use the Help Sheet from your teacher to find out the correct answers.

Element	My guess for the symbol	Correct answer
carbon		
sulphur		
nitrogen		
oxygen		
fluorine		
phosphorus		

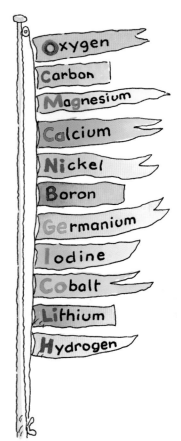

a Copy and complete the sentence to give a simple rule for writing the symbols for elements.
For some elements the symbol is the . . . . of the name of the element.

Now find the symbols for:

b calcium   **c** chlorine   **d** chromium.

The names of these elements all begin with the same letter.
The symbols use a second letter from the name too.
We always write the second letter as a small letter.

e The symbols for copper, iron and sodium do not fit in with these rules. Where do you think their symbols come from?

What **A**re **Th**e **S**ymbol **Ru**les? (Find the elements!)

Atoms can join together. They make **molecules**.

2 oxygen atoms can join together. They make an oxygen **molecule**.

A carbon atom can join to 2 oxygen atoms. They make a **molecule** of carbon dioxide. Carbon dioxide is a **compound**. It has *different* atoms joined together.
Notice that you can have molecules of an *element* or molecules of a *compound*.

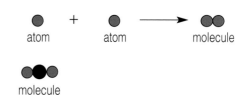

Spotting elements and compounds

Look at these diagrams of atoms and molecules.
Match the answers to the boxes.

| **f** | **g** | **h** | **i** | **j** |

i) Atoms of one element.
ii) Molecules of one element.
iii) Molecules of one compound.
iv) A mixture of 2 elements.
v) A mixture of 2 compounds.

Now draw your own boxes to show these:

k A mixture of 1 element and 1 compound.
l A mixture of 3 compounds.

International science

All the symbols for elements are international. Maybe you can't
understand the language. But you can spot the chemical elements!
How's your Russian?

Use books or ROMs to find out about one of the elements
mentioned in this Russian book, plus one other element.
Make a poster to show what you have found out.

...I don't think he's an element!

ANALOGIJA MED DVODIMENZIONALNO
RAZPOREDITVIJO IGRALNIH KART IN
PERIODNIM SISTEMOM ELEMENTOV

Opredelitev problema:

Sredi 19. stoletja so kemiki opazili, da se kemijske
lastnosti elementov periodično spreminjajo z naraščajočo
(relativno) atomsko maso. Elemente so začeli razvrščati
po kemijski sorodnosti. Tako je Wolfgang Döbereiner že
leta 1829 razvrstil elemente po sorodnosti v trojke (tria-
de).

Li	Cl
Na	Br
K	I

podobna tališča, gostote,
podobna kemijska reaktivnost

---

1 Copy and complete the table:

Symbol	Name
C	
Cu	
	oxygen
N	
Ca	
	iron
	sodium
	chlorine
Mg	
S	

2 a) Write down the names of 2 elements
in each case which are:
i) solids   ii) liquids   iii) gases.
b) Draw diagrams to show how the
particles of the elements are arranged in
solids, liquids and gases.

3 The table shows the approximate
percentages of different elements in rocks of
the Earth's crust.

Element	Percentage
oxygen	48
silicon	26
aluminium	8
iron	5
calcium	4
sodium	3
potassium	2
magnesium	2
other	2

Draw *either* a pie-chart *or* a bar-chart of
this information.

Things to do

Classifying elements

Learn about:
● metals and non-metals
● properties of elements
● uses of elements

▶ Look at the objects in the photograph:
Divide them into 2 groups:

● Those you think are made from *elements*.

● Those you think are made from *compounds*.

Materials can be sorted into groups. We say they can be **classified**.
You can do this by testing some *properties* of the materials.

▶ Look at the properties opposite:
How could you test for each of these?
What equipment would you use? What would you do?

Some of these tests could be used to put *elements* into groups.

There are more than 100 elements.
Some are hard to classify. Most can be put into 2 groups.
The groups are **metals** and **non-metals**.

strength   hardness   density
conducting electricity
melting point   boiling point
conducting heat

▶ Copy out the table.
Make it at least 10 cm long and 10 cm wide.
Fill in your ideas about the properties of metals and non-metals.
(You could check these with your teacher.)

Property	Metal	Non-metal
appearance		
strength		
hardness		
density		
melting and boiling points		
does it conduct heat?		
does it conduct electricity?		

Are you sure it's metal?

Metal or non-metal?

Test the elements your teacher will give you.
Decide whether each element is a metal or non-metal.
Check your plans with your teacher before you start.

The uses of an element depend on its properties.

a Why is aluminium used to make aircraft bodies?

c Why is gold used to make jewellery?

b Why is copper used to make saucepans?

Choose the cable

Liftum Ltd is a new company. It makes cables to carry cable cars. Each cable car seats 4 people.

Can you recommend a material to use for making the cables?
You could use copper, iron or aluminium.
Which would be best?

In your group, discuss which factors you will need to consider.
- Write out your list of factors.
- Which material do you *think* is the best?
- Why have you chosen this one? Are you happy with your choice?

Now *plan* some tests on the 3 materials to see which *is* best.

Before starting your practical work get your teacher to check your planning.

How can you make your results reliable?

It would be important for Liftum to know how much these materials cost.
Which do you think is the cheapest? Why?
Try to find out the cost of each one.

1 Write down the names and symbols of:
a) 5 metals   b) 5 non-metals.

2 Make a list of words which describe metals. Then make a list of words which describe non-metals.

3 Remi has found a lump of black solid. It is light, breaks easily and doesn't conduct electricity.
Is it a metal or non-metal?

4 Some objects can be made of metal or plastic. Discuss the advantages and disadvantages of metal or plastic for each of the following:
a) ruler   b) window frame
c) spoon   d) bucket.

5 Look around the room.
a) Name 4 objects made of metal.
b) Which metals are they made from?
c) Why are the metals used to make the objects?

Things to do

The periodic table

Learn about:
- looking up data on elements
- the periodic table

What do you remember about **elements?**

Elements are substances which *cannot be broken down into anything simpler*.
An element has *only one type of atom*.

You've met lots of elements already.
Which of the elements in the box:

a has the symbol H?

b has the symbol S?

c has the symbol Cu?

d turns blue with starch?

e is about 21% of the air?

f What does *classifying* mean?

Elements
copper
hydrogen
iodine
oxygen
sulphur

Scientists classify the elements in the **periodic table**.
But how are the elements arranged?
Let's look for patterns.

The elements here are sodium, gold, copper and chlorine. Can you identify each one?

How are books classified in your school library?

Grouping elements

▶ Look at this information about elements.

Element	Appearance	Reaction with cold water	Other information
sodium	silver-grey solid	reacts violently	conducts heat and electricity
silver	shiny silver solid	no reaction	conducts heat and electricity
helium	colourless gas	no reaction	very unreactive, doesn't conduct
lithium	silver-grey solid	reacts very quickly	conducts heat and electricity
copper	shiny pink-brown solid	no reaction	conducts heat and electricity
chlorine	green-yellow gas	dissolves	reactive, doesn't conduct
argon	colourless gas	no reaction	very unreactive, doesn't conduct
fluorine	pale yellow gas	dissolves	very reactive, doesn't conduct
gold	shiny gold solid	no reaction	conducts heat and electricity
potassium	silver-grey solid	reacts very violently	conducts heat and electricity
neon	colourless gas	no reaction	very unreactive, doesn't conduct
bromine	red-brown liquid	dissolves	reactive, doesn't conduct

- Finding out the melting point and boiling point of each element in the table above.

Make a data card for each element.
Use your cards to match up similar elements.
You can move the cards around to get the best match.

- Which elements have you grouped together? Why?

- Within each of your groups, put the elements in order.

- Explain your order.

DATA CARD

Element _____
Appearance _____

Melting point _____
Boiling point _____
Reactivity _____

Mendeleev was a Russian scientist. In 1869 he made a pattern of elements. He put the elements in **groups**. This pattern is called the **periodic table**.

Your teacher will show you a copy of the **periodic table**. The columns of elements are called **groups**. The rows of elements are called **periods**. Are *your* groups the same as those in the periodic table?

g Which elements are in the same group as chlorine?

h Name 2 elements in the same period as chlorine.

i Where are the *metals* in the periodic table?

j Within each group, where are the *most reactive* metals?

Mendeleev (1834–1907)

Your teacher may show you some metals reacting. How do sodium and magnesium react with water?

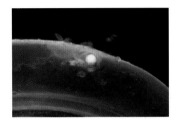
sodium and cold water

magnesium and cold water

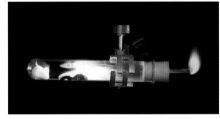
magnesium and steam

k Which is more reactive, sodium or magnesium?

▶ Look at a copy of the full periodic table.

l Write the symbol and name of the most reactive metal. (You can use your answers to **i, j** and **k**, to help you answer this.)

Video highlights

The Spotlight Video Company needs your help.
The company wants to make a new video.
It will be about elements and the periodic table.

- The video will be for 13-year-old pupils.
- It should be about 5 minutes long.
- It should be interesting and exciting. (It *might* be funny!)
- It should explain about elements and the periodic table.

Your group should write a script for the video.
What will be seen on the screen?
Explain what filming will be needed.
Who would you like to present your video? Money is no object!

1 Copy and complete:
a) Elements have only one type of . . . .
b) Elements are classified in the . . . . table.
c) The columns of elements are called . . . .
d) The rows of elements are called . . . .

2 Find out about other scientists who tried to group elements e.g. Newlands.

3 Choose one group of elements. Use books to find out about the group. Make a wall poster about the elements.

Things to do

Making compounds

▶ Match each of the elements in the box with one of the descriptions **a** to **f**.

a an element with the symbol U

b a metal used to make cooking foil

c a metal used for electrical wires

d a metal which is a liquid

e an element with the symbol K

f a non-metal

- potassium
- uranium
- copper
- mercury
- carbon
- aluminium

You have tested properties of many elements. For example, strength and hardness. These are **physical properties**.

The **chemical properties** of an element are also important. Does the element change easily into a new substance? Does it **react?**

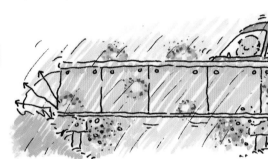

Remember! The use of a material depends on its properties.

Don't make a bridge from a metal which reacts with water! Which element is best to use? Why?

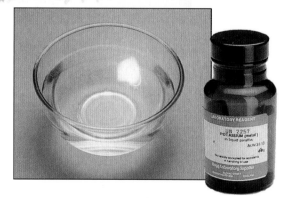

Potassium is a metal. It reacts very violently with water. Potassium is stored under oil. Why do you think this is?

Reacting metals

Do metals react with oxygen in the air?

- Predict the product formed when magnesium is heated in air.
- Get a small piece of magnesium ribbon from your teacher. Hold it at arm's length in some tongs. Then move it into a Bunsen burner flame (air-hole just open). ***Do not look directly at it***. What happens?

eye protection

- Do the experiment again using copper foil rather than magnesium. What happens? Which is the more reactive, magnesium or copper?

When metals react with oxygen they make new substances.
These are called **oxides**.

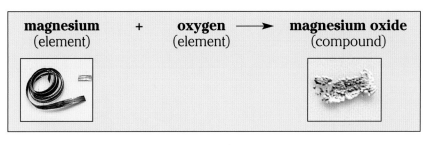

magnesium + oxygen ⟶ magnesium oxide
(element)   (element)   (compound)

Will the magnesium oxide weigh more than the magnesium you started with? Why?

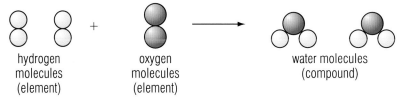

copper + oxygen ⟶ copper oxide
(element)   (element)   (compound)

Non-metals can react with oxygen too.
They also make **oxides**.
Water is made from the **elements** hydrogen and oxygen.
The **elements** combine to make a **compound**.
You can use pictures to show this:

hydrogen   oxygen   water molecules
molecules   molecules   (compound)
(element)   (element)

Why can't you **see** water molecules?

g How many atoms of hydrogen make up each hydrogen molecule?

h How many atoms make up each water molecule?

i What is the proper *chemical* name for water?

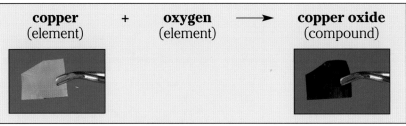

1 Copy the diagram.
Put the correct words in the empty boxes.

metals compounds elements non-metals

pure substances

very reactive unreactive
reactive

2 Use books or ROMs to find out when different metals were discovered. Make a time-chart to display in the laboratory. Look for a pattern between the discovery dates and the metal's reactivity.

3 Write down the word equations for the following reactions:
a) aluminium and oxygen.
b) zinc and oxygen.

4 Water (H_2O) is a compound.
We can break it down into its elements by passing electricity through it.
a) Write a word equation for the breakdown of water.
b) Draw the molecules involved in the reaction in part a).

5 Ammonia (NH_3) is a compound.
a) How many atoms of nitrogen are in each ammonia molecule?
b) How many atoms of hydrogen are in each ammonia molecule?

Things to do

Questions

1 What do you think each of the following words means?
Write no more than 2 lines for each.
a) atom b) molecule c) element d) compound.

2

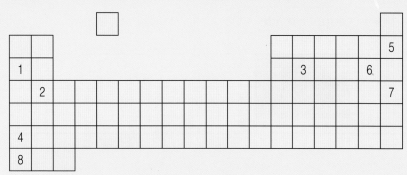

In the outline of the periodic table, the numbers represent elements.
Give the **numbers** of:
a) 3 elements in the same period
b) 3 elements in the same group
c) 2 metals with similar properties
d) 2 non-metals with similar properties
e) the most reactive metal.

Newlands   Döbereiner

Mendeleev

3 Name an element in each case which:
a) is a gas at room temperature and made of molecules
b) is always present in sulphides
c) is a non-metal and a liquid at room temperature
d) is a metal and a liquid at room temperature
e) has the symbol N
f) is a more reactive metal than sodium
g) is present in carbon dioxide
h) is the non-metal present in iron chloride.

This liquid non-metal has a hazard warning

4 Metals can be mixed together to make **alloys**.
Find out the metals in each one of these alloys:
a) brass
b) solder
c) bronze
d) Duralumin.
Write down one use for each of the alloys.

A stainless steel sink is an alloy of iron

5 Hydrogen burns in chlorine gas to produce hydrogen chloride.
a) Write a word equation for this reaction.
b) Under your word equation, show which substances
   are elements and which is a compound.

6 Plan an investigation to put the following materials into an
order of hardness:

copper   iron   zinc   steel

Copper is used to make pipes

Compounds and mixtures

Everything around you is an element, a mixture or a compound.
You are a combination of elements, mixtures and compounds.
The elements are the building blocks.
There are 92 natural ones. How many can you name?

In this unit you will be looking at lots of elements.
Some will be in mixtures . . . . Some will be in compounds . . . .

In this unit:

Compounds

Remember that compounds form when 2 or more different atoms join together. The symbols for elements can be used to write a **formula** for a compound. For example,

CuO is copper oxide (ox**ide** when O is in a compound)
LiCl is lithium chloride (chlor**ide** when Cl is in a compound)

▶ What do you think are the names for the following compounds? Write them down.

a KCl **b** CaO **c** MgO **d** NaCl

Some compounds have more complicated formulae. Look at:

$CuCl_2$

This is copper chloride. The compound has 1 copper atom and 2 chlorine atoms. How can you tell this from the formula?

▶ Copy and complete the following table. The first one has been done for you.

Name	Formula	Number of each type of atom
carbon dioxide	CO_2	1 carbon, 2 oxygen
sodium fluoride		1 sodium, 1 fluorine
	$MgCl_2$	
	$AlCl_3$	
lithium oxide	Li_2O	

Compounds whose name ends in the letters **-ide** are usually '2-element' compounds. Here are some examples:

Compound	Elements in the compound	Formula
zinc bromide	zinc and bromine	$ZnBr_2$
aluminium iodide	aluminium and iodine	AlI_3
magnesium sulphide	magnesium and sulphur	MgS

▶ Name the compound formed between:

e calcium and iodine

f lead and bromine

g copper and sulphur.

The exception to the '-ide' rule are compounds called hydroxides. For example, the formula of sodium hydroxide is NaOH.

h Which elements make up sodium hydroxide?

black copper oxide

copper being heated in air

The copper combines with oxygen in the air to make copper oxide

The **formula** of a substance tells us how many of each type of atom are present, e.g. the formula of water is H_2O, so there are twice as many H atoms as there are O atoms.

Now we've joined together, I've changed my name from sulphur to sulphide.

Some compounds contain more than 2 types of atom.

The ones you need to know about often contain oxygen in '3-element' compounds.

We can usually recognise these compounds as their names end in **-ate**.

Here are some examples:

Compound	Elements in compound	Formula
copper sulphate	copper, sulphur and oxygen	$CuSO_4$
calcium carbonate	calcium, carbon and oxygen	$CaCO_3$
lithium nitrate	lithium, nitrogen and oxygen	$LiNO_3$

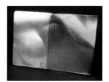 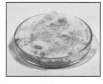

copper   sulphur   oxygen

These elements make up the compound called copper sulphate

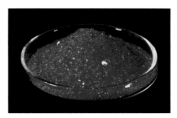

copper sulphate

▶ Which elements make up these compounds:

i sodium carbonate?

j magnesium sulphate?

Elements to compounds

Watch your teacher heat some zinc powder and sulphur powder in a fume cupboard.

- Describe the elements before the reaction.
- What do you see as the reaction takes place?
- Write a word equation for the reaction.
- How can you tell that a compound has been made?

a **mixture** of zinc and sulphur

heat

 + ⟶

Symbol equation: Zn   +   S   ⟶   ZnS

Things to do

1 Use your own colours for atoms.
Draw 4 different boxes to show:
a) a mixture of 3 elements,
b) a pure compound,
c) a pure element,
d) a mixture of 2 compounds.

2 Draw a cartoon to show zinc combining with bromine to make zinc bromide.
(Hint: check the formula from page 72.)

3 A mixture can contain 2 elements.
A compound can also contain 2 elements.
What are the differences between mixtures and compounds?
Write down your ideas.

4 Copy and complete this table:

Compound	Elements present
	copper, nitrogen oxygen
	sodium, carbon, oxygen
	tin, bromine
	zinc, iodine
potassium chlorate	
silver chloride	

73

Mixed or fixed?

In a compound, two or more elements join together.
But how much do you know about **mixtures**?

▶ Look at these ideas about mixtures and compounds.

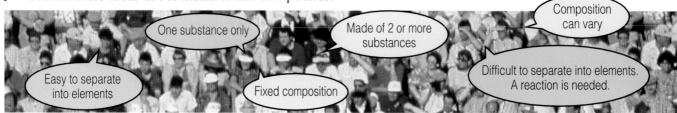

One substance only

Made of 2 or more substances

Composition can vary

Easy to separate into elements

Fixed composition

Difficult to separate into elements. A reaction is needed.

Draw a table like this:
Write each idea in the correct column.
(You should have 3 in each.)

Mixture	Compound

Lots of natural substances are mixtures.
Some are mixtures of elements. Some are mixtures of compounds.
Often the mixtures must be separated before we can use them.

I'm not keen on the local water. Are you?

I ♥ SCIENCE

You should have separated your air!

I'm having a few problems with the car!

CRUDE OIL

a How could you improve the water?

b Which gas is in Lisa's balloon?

c How could you get the petrol?

▶ Look at these photos. They are about mixtures.
Can you explain the separations?

g How can you get the cream from the milk?

This is a hot country.
You can get salt from sea-water.

d How is the salt separated from the sea?

Salt is sodium chloride NaCl.

e Will it be easy to get chlorine from salt? Explain your answer.

This is a filter coffee maker.

f How does it work?

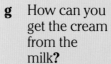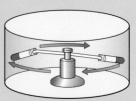

This is a centrifuge. It is used to separate blood cells from plasma.

h How do you think it works?

Making a mixture and a compound

Do you know the symbols for iron and sulphur?
Your teacher will show you how to make a mixture
and then a compound from these elements

i What is the compound called?

j How do you know a ***compound*** has been made?

k How is the compound different from the elements
and the mixture?

l Can you write a formula for the mixture? Explain your answer.

Iron is attracted to a magnet

Sulphur is not attracted

Compounds react too!

Follow the instructions opposite to see compounds reacting
together.

m Write down everything you see.

n How do you know there is a reaction?

The symbol equation is:
$CuO + H_2SO_4 \longrightarrow CuSO_4 + H_2O$

o Copy and complete the word equation:
. . . . + sulphuric acid $\longrightarrow$ copper sulphate + ...

Half fill a test-tube with dilute sulphuric acid.
Add one spatula measure of copper oxide.
Stir the liquid.
Leave the tube in your
rack for 5 minutes.

acid irritant
eye protection
copper oxide is harmful

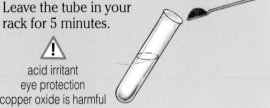

Discussing elements and compounds

▶ Look at each statement below.
Is it a description of an element? Is it possible to tell?
Write down your ideas. Use a code for each answer: ✓ for *element*
? for *not possible to tell*
✗ for *non-element.*

p made of only one type of atom

q is called water

r looks shiny

s made of atoms

t gives off carbon dioxide when
heated

u made of molecules

v has the symbol Sn

w has the symbol O_2

x has the formula CO

y has the formula FeS

This is a model. It shows how different
atoms are joined together in a large
molecule of a compound.

▶ Discuss the ideas with others in your group. Do you all agree?

---

1 Copy and complete.
Choose a word from the box:

> mixture   easy
> hard   compound

a) A . . . . has fixed composition.

b) In a . . . . the elements or compounds
do not combine.

c) It is . . . . to separate a compound
into its elements.

d) Usually a mixture is . . . . to separate.

2 Two elements join together.
How can you tell a reaction has taken place?
Write down 3 ways.

3 Say whether each of these substances
is an element (E), mixture (M) or
compound (C).

> magnesium   air   hydrogen   water
>   iron oxide   salty water   chlorine

Are any substances difficult to classify?
Explain.

Things to do

Separating mixtures

Learn about:
- separating mixtures
- separating gases from air
- mixtures and pure substances

We can divide all substances into 2 groups:

pure substances   or   **mixtures**

A **pure substance** contains just one element or just one compound.
A **mixture** has two (or more) elements or compounds in it.
These are just *mixed* together. They are not joined together.
If you have a **mixture** you can separate it into pure substances.
Sometimes this is easy to do. Sometimes it is hard to do.

Try out these exercises. Make sure you write down your answers carefully.
There's lots to do here!

Look for clues

Here are 2
elements, one
mixture of them
and one compound
of them.

a   Name the non-metal element.

b   Which is the mixture of iron and
    sulphur**?**

c   Which is the compound**?**

d   How could you get iron from the
    mixture**?**

Now 2 more difficult questions:

e   How could you get sulphur from the
    mixture**?** You must use a different
    method to **d**!

f   What is the compound called**?**

Colour me in

Get a copy of the periodic table.
Think about the elements you
know.
Colour the metals in blue.
Colour the non-metals in red.

Find the element

Make up your own word search
for elements.
Write the names of the elements
horizontally (across) or
vertically (down).
Hide at least 6 elements in the
middle of other letters.
Keep a record of your answers.
Let someone else in your group try the word search.

Separating gases in the air

As we saw in Book 7, air is a mixture of gases. We can separate
the gases using a **fractionating column**. This is a huge tower.

The air is first filtered to remove particles of dust.
It is then cooled so that all the gases in it change to
liquids. Inside the fractionating column the liquid
air is slowly warmed. As each substance in the
liquid air reaches its boiling point, it evaporates.
The gases can then be collected separately.
This is called **fractional distillation**.

The boiling points of the main gases in air are shown
in the table opposite:

Use this table to explain how oxygen is separated from liquid air.

Gas	Boiling point (°C)	Approx. % in air
Nitrogen	−196	78
Argon	−186	1
Oxygen	−183	21

A complete wreck

Read this extract from the *Dove Times*:

Experts were worried last night. A strange mixture was found on Dove Beach. A local man thinks it is from the ship *The Red Lady*. It was wrecked a few months ago down the coast. It was made of iron. It carried a cargo of salt and limestone. But sand from the beach may be in the mixture too.

Plan an investigation to get the pure substances from the mixture.
Try to get pure iron, salt and sand.
Show your plan to your teacher.
If it is safe, you can carry out your investigation.

Needing water

The crew from *The Red Lady* might be in trouble!
In order to survive, they need pure water.
The apparatus opposite will help!

g Copy the diagram. Label the apparatus.

h What is the process called?

i Explain the process. Use the words: evaporate, condense.

j What does the thermometer do?

k How do you know the water is pure?

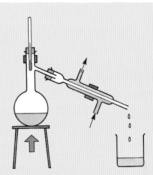

This apparatus gets pure water from sea-water.

1 Copy and complete. Use the words in the box. Some words can be used more than once.

elements	compounds
reactive	unreactive	non-metals
mixture	metals

a) Pure substances are . . . . or . . . .
b) Elements can be . . . . or . . . .
c) Some metals are . . . . e.g. Mg.
d) Some metals are . . . . e.g. Cu.
e) When elements combine they make . . . . e.g. CuO.
f) In a . . . . the elements or compounds are not joined together.

2 Imagine you have discovered a new element.
Describe 4 tests you could do to see if it is a metal or a non-metal.

3 Do some research to find out more about one of the gases we find in air.
You can use reference books, videos, ROMs or the internet.
Present your findings on a poster or as a leaflet to share with the rest of your class.
(Hint: your work in Unit 7F may help you here.)

Things to do

Chemistry at Work

Elements made into compounds

Sulphur
Sulphur is used in the first step in making sulphuric acid. Molten sulphur is sprayed into a furnace and burned in a blast of hot air.

a Which gas in the air does the sulphur react with?

b Write a word equation to show what happens when sulphur burns in air.

c Why do the gases given off from a factory making sulphuric acid have to be carefully monitored?

Sulphuric acid on the move

Look at the table below showing the uses of sulphuric acid:

Uses of sulphuric acid	Percentage used (%)
Making new chemicals	25.7
Paints and pigments	21.6
Detergents and soaps	13.3
Fertilisers	11.2
Plastics	7.0
Fibres	5.3
Dyes	2.8
Other uses	

d Work out the percentage of sulphuric acid that should be under the heading 'Other uses' in the table above.

e Use a computer to display the uses of sulphuric acid in a pie-chart.

Magnesium
Magnesium is used in fireworks and flares.

f What is the colour of the flame produced when magnesium burns?

g Write a word equation for magnesium burning.

h The makers of fireworks want magnesium to burn more quickly than a piece of magnesium ribbon. What can they do to the magnesium to make this happen?

Mixtures

Helium and oxygen

Did you know that helium is used by deep-sea divers? It is mixed with oxygen gas for divers to breathe in. This is safer than breathing in normal air which contains nitrogen. The nitrogen can dissolve in a diver's blood and can form a bubble as the diver rises to the surface. This can kill the diver.

i Which gas do you think is more soluble in blood, nitrogen or helium? Why do you think this?

j What do divers call it when a bubble of gas is formed in their circulatory system?

k What happens to the voice of a diver breathing in a mixture of oxygen and helium?

l Find out about and explain another use of helium.

Divers use a mixture of helium and oxygen

Alloys

An alloy is a *mixture of metals*.
Alloys are made to improve the properties of a metal.
For example, steel is an alloy of iron.
It has traces of carbon mixed with the iron, but can also have other metals added.
Tungsten mixed in makes the steel very hard and is used to make cutting tools.
Cobalt and nickel are added to make stainless steel, which does not rust.

m Which type of steel would you use to make the sharp edge of a lathe in your technology department?

n Name some objects that are made from stainless steel.

o Why do you think cars aren't made from stainless steel?

p The atoms of a pure metal are all the same size and packed closely together in layers.
Explain why adding atoms of a different size might make it more difficult to bend the metal.

Using a lathe

Alloys are used to make aeroplanes and coins

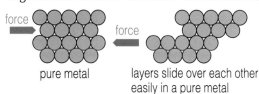
force   force

pure metal   layers slide over each other easily in a pure metal

q Find out about the alloys used to make aeroplanes or the alloys used to make coins.

Developing Chemistry

Learn about:
- investigating unknown liquids
- ideas changing in time

What do you remember about melting points and boiling points?
Think about these temperatures:

a What is the melting point of ice?

b What is the freezing point of water?

c What is the boiling point of water?

d What happens if the ice or water is not pure?
Will the freezing point or boiling point be different?
Write down your ideas.

e Is the boiling point of a substance *always* higher than the melting point?
Explain your answer.

Does salty water boil at the same temperature as pure water?

Pure or impure?

Pure elements and pure compounds boil at one particular temperature. They will also melt at a specific temperature.
So the boiling point and melting point of a pure element or compound are *characteristic* of that substance.
They are always the same for the pure substance.

However, mixtures melt or boil over a range of temperatures – not sharply at one particular temperature.
The temperature at which a mixture starts to melt or boil will also depend on its composition.

Look at the enquiry opposite:
Plan what you could do.
Discuss your idea with your teacher.
You may be able to carry out your plan.

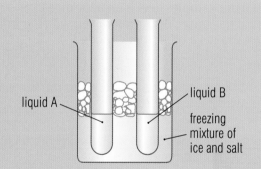

liquid A

liquid B

freezing mixture of ice and salt

Find out which liquid is pure and which is a mixture

Ideas about elements, mixtures and compounds

The famous Greek philosopher **Aristotle** had a theory about the basic building blocks of all substances.

He believed that everything was made up from different combinations of:

- earth,
- air,
- fire, and
- water.

These ideas were put forward around 350 BC, but were not developed much further for centuries.

Nobody could come up with any better way to explain the different properties of materials.

ARISTOTELL STAGIRITAE

The alchemists

It wasn't until about 775 AD that Aristotle's ideas were eventually refined. An Islamic **alchemist** called **Jabir Ibn Haiyan** accepted the Greek ideas. He liked the theory but needed to modify it to explain his own observations and experiments. The alchemists were very interested in gold. One of their aims was to turn other metals into gold. In their efforts to achieve this, they carried out thousands of experiments. The Islamic alchemists translated the works of the ancient Greeks and used their ideas to guide their experiments. They were the first to see the important link between theory and experiment. In fact, Jabir is sometimes called the ***Father of Chemistry***.

According to Greek theory, minerals and metals are mixtures of earth (on the way to becoming fire) and water (on the way to becoming air). The 'earthy' bit dominates in minerals, and the 'watery' bit dominates in metals.

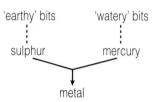

Jabir suggested a slightly different theory.
He said that deep underground the 'earthy' bits are changed into sulphur (a brittle, yellow non-metal). The 'watery' bits are changed into mercury (the silvery, liquid metal you've seen in thermometers).
Then the sulphur and the mercury combine to form one of 6 metals that they knew about.
These were iron, lead, copper, tin, silver, and, of course, gold. The metals were different because they contained different amounts of sulphur and mercury. The sulphur and mercury might also be impure in some metals.
Jabir described gold as the perfect combination of pure sulphur and mercury. With his theory, it now seemed to make sense to think that you could treat one metal and convert it into another. If you came up with the perfect combination, you hit the jackpot and made gold!

Equipment used by the alchemists

The theory inspired alchemists for hundreds of years to carry on the search for a way to change a cheap metal, such as lead, into gold.
But they never did succeed!

1 Copy and complete:
a) . . . . elements and compounds melt and boil at particular temperatures.
b) Mixtures do not . . . . or . . . . at fixed temperatures.
c) The melting point and freezing point are always the . . . . for a particular substance.

2 What did the Greeks believe about the way matter is made up?

3 The Greeks also had ideas about disease and illness, based on their ideas about matter. Imagine you are an ancient Greek doctor. Your patient has a very high fever. Explain to the patient what you think is wrong and suggest a treatment.

4 Why do you think that Jabir is called the ***Father of Chemistry***?
Explain why Jabir might have had his theory about metals, after looking at a piece of gold.

5 Why did the alchemists never manage to change lead into gold?

6 Jabir made many important discoveries. Do some research and present a poster on Jabir's contribution to chemistry.
(You will find other references about him under the name Geber. This was how his name was translated when European scientists discovered his work centuries later.)

Things to do

Questions

1 Crude oil is a mixture of compounds.
A sample of crude oil was found to contain the following:

Substance	Amount in crude oil (%)
refinery gas	0.2
gasoline	30.0
naphtha	7.0
kerosine	10.0
diesel oil	30.0
fuel oil	20.0
lubricating oil	2.0
bitumen	0.8

Put this information in the form of a bar-chart.

2 Use books or ROMs to find out which elements make up these compounds:

a) PVC   b) paper   c) candle wax   d) Teflon

3 a) Which boxes contain the elements, mixtures or compounds?
Describe each box as fully as you can.

    1   **2**   **3**   **4**   **5**

b) Draw a box that shows two elements reacting together to form a compound.

4 Look at Ann's science homework:
From Ann's work:
a) Write the names of 2 elements.
b) Write the name of 1 compound.
c) Was there a reaction? How do you know?
d) Write a word equation.

> I lit a spill. I put it into a tube of hydrogen gas. The gas reacted with oxygen in the air. I saw a flash. I heard a 'pop' sound. I made some water in the tube.

5 Here are some formulae for compounds. See if you can work out their names.

a) MgO c) HCl e) CO_2

b) $CuCl_2$ d) FeS f) H_2O

These are a little more tricky!

g) KOH h) $ZnCO_3$ i) $CaSO_4$

Rocks and weathering **8G**

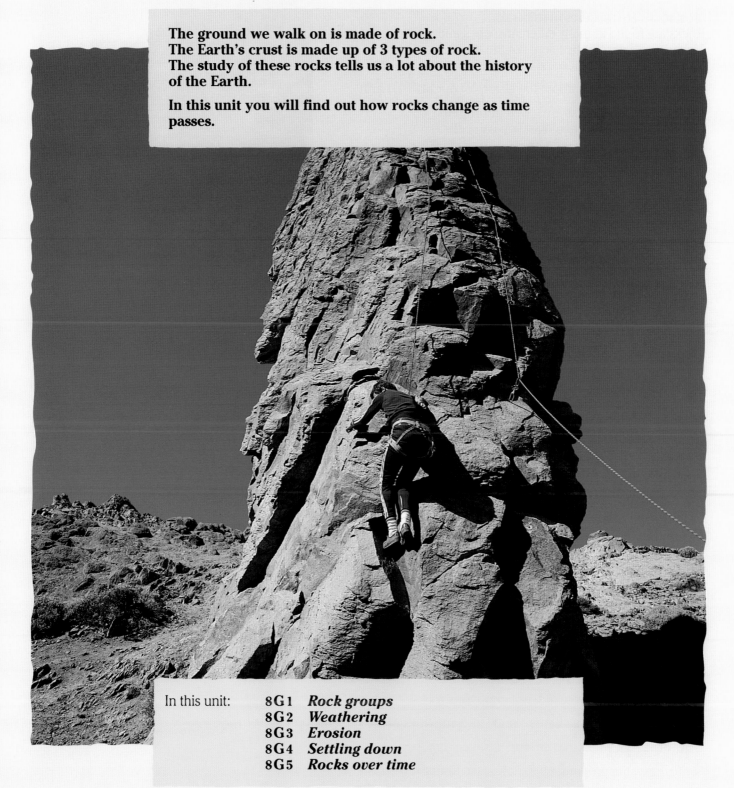

The ground we walk on is made of rock.
The Earth's crust is made up of 3 types of rock.
The study of these rocks tells us a lot about the history of the Earth.

In this unit you will find out how rocks change as time passes.

83

Rock groups

Most rocks are mixtures of minerals.
Rocks are found in many different shapes and sizes.
Maybe you have climbed some of the larger ones ...
or maybe you have collected some.

Rocks can tell us a lot about the history of our planet, Earth.

People who study rocks are called **geologists**.
Let's see how good a geologist you can be!

Testing rocks

Carry out these tests on each rock sample.

Your teacher will give you some Rock Data Cards.
Write down the results for each rock on a new Rock Data Card.

Rock test 1 – What does the rock look like?

Use a magnifying glass to observe the rock carefully.
● What colour is it?   ● Is it rough or smooth?
● Is it shiny or dull?   ● Can you see any crystals or grains?

Rock test 2 – Is the rock hard?

Try to scratch the rock.
Rocks which can be scratched by a fingernail are called **very soft**.
Rocks which can be scratched by an iron nail are called **soft**.
Rocks which can be scratched by a steel knife are called **hard**.
Rocks which cannot be scratched by a steel knife are called **very hard**.

Rock test 3 – Does the rock break easily?

Wrap the sample in a cloth. Put it on the floor.
Lower your heel onto the rock. Push down.
Does the rock break?

Rock test 4 – Does the rock soak up water?

Put the sample on a watch glass.
Use a pipette to drop water on to it.
What happens to the water?

Rocks that soak up water are called **porous** rocks.

Look at your Rock Data Cards. Study the properties of the samples.

Now divide the rocks into groups. List your groups.

Write down the properties you have used to make groups.

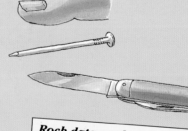

Rock data card
Sample number
Rock test 1

Rock test 2

Rock test 3

Rock test 4

Other information

Rocks contain different particles or grains.
Some grains fit together well. They have an **interlocking texture**.
But in some rocks the grains do not fit so well.
We say that they have a **non-interlocking texture**.

Think about the rocks which soak up water.
Where do you think the water goes?

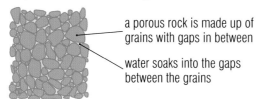

a porous rock is made up of grains with gaps in between

water soaks into the gaps between the grains

Another way of grouping rocks is by the way they were made.

There are 3 main types of rock.
The 3 types were formed in different ways.

Porosity

Measure how much water a rock soaks up.
Compare sandstone and granite.
Which is more porous?
Use some building blocks to model the different structures of sandstone and granite.

Granite (igneous)

Igneous rocks
These form when melted ('molten') substances cool.
These rocks are usually hard.
They are made of crystals.
Granite is an igneous rock.

Sandstone (sedimentary)

Sedimentary rocks
These form in layers. They are made when substances settle out in water.
Sometimes they contain fossils.
These rocks are usually soft, but not always.
Sandstone and **limestone** are sedimentary rocks.

Metamorphic rocks
These are made when rocks are heated and/or put under great pressure.
They are usually very hard.
Marble is a metamorphic rock formed from limestone.

Marble (metamorphic)

Limestone (sedimentary)

---

1 Copy and complete the following:
a) There are 3 main types of rock: . . . . , ... and . . . .
b) . . . . rocks form when hot liquids cool and become solid.
c) . . . . rocks form as substances settle in layers.
d) . . . . rocks form when other rocks are heated and pushed together.
e) Rocks that soak up water are called . . . . rocks

2 Use your rock test findings, books and computer to decide if your rock samples are igneous, sedimentary or metamorphic.

3 Some types of rock are used in buildings.
a) What are the ideal properties of a rock used for building?
b) Find out what types of rock are used for your local buildings.
Draw a poster about this to display in your local library.

4 Can you explain how the rocks in this photograph have changed?

Things to do

Weathering

Learn about:
● different types of weathering

Rocks don't stay the same forever.
They slowly crumble away.

▶ Look at these photographs and say what you think has caused the rocks to change in each case.

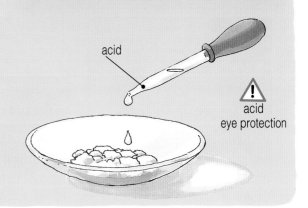

The process that makes rocks crumble is called **weathering**. Weathering can be caused by rocks reacting with water or substances dissolved in water and by changes of temperature.

Acid attack

Do you remember looking at how acid rainwater affected limestone in Unit 7F?

Carry out an investigation to see if acid affects other rocks in the same way.
Add a few drops of acid to chalk. Write down what you see.

Repeat your investigation with granite, sandstone and marble chips.

Leave the granite in the acid for a few weeks.
You might be able to record any changes using a digital camera.
Explain your observations.

acid

⚠ acid
eye protection

Frost damage

Water can get into cracks in rocks.

If the water freezes, it turns to ice. But ice takes up more space than water. So the ice can split the rock into smaller pieces after this has taken place many times. This is called frost damage.

Your teacher will fill a small glass bottle with water and screw the top on tightly.

The bottle is then put into a strong plastic bag, which is tied and put into a freezer.

In the next lesson your teacher will show you what has happened to the bottle.

Rocks can also be broken down by changes in temperature. Each mineral in a rock will expand and contract at different rates as it gets hot or cools down.
This results in stress within the rock. This eventually causes it to crack.

▶ Explain why the minerals in rock expand and contract when the temperature changes. Use the word 'particles' in your answer.

The rocks in a desert are regularly exposed to extreme temperature changes

Investigating weathering

Look at rocks around your school for signs of weathering. You can also look at bricks and other building materials.

- Are the weathered rocks soft?
  Scratch them with an iron nail to find out.
- What colour are they?
  Is their colour different from that of unchanged rock?
- Are there any cracks in their surface? What are the cracks like?

- What do you think caused the weathering in each case?
- What types of rock crumble most easily?
- What types of rock last the longest?

Look to see how mosses, lichens and other plants can change rocks. Examine the rock underneath these plants and then carefully replace them.

▶ Find out about different types of weathering:
- physical weathering
- chemical weathering
- biological weathering.

Things to do

1 Copy and complete the following:
The process that makes rocks crumble is called . . . . Soft rocks crumble more easily than . . . . rocks.
Rain can weather rocks because it is . . . . When water gets into cracks it can . . . . to form . . . . which takes up more space and so it can split the rock into smaller pieces. Changes in . . . . can also cause rocks to weather.

2 Look at this diagram. Write an explanation of how loose scree forms.

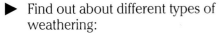
loose scree

3 Visit your nearest churchyard or cemetery. But never go on your own. Always take a friend or an adult with you.
Look carefully at the different types of gravestone. Some will have weathered more than others.
What are the earliest dates that you can read? These will be on the hardest rocks. Try to name the different types of rock and note the earliest date on each.

8G3 *Erosion*

Weathering makes rocks crumble into smaller pieces.

These pieces are then carried away by other things, e.g. wind, and so the rock wears away.

This wearing away of rocks is called **erosion**.

CRUMBLE COTTAGE
LAND-SLIDE WRECKS CLIFF HOMES

The Robinsons are moving house, because their house is on the move.
The garden is not what it was, in fact it's nearly all gone.

The floors are tilting and the walls are cracking.
The Robinsons, who have lived there for 30 years, said 'We are so sad. When we

bought the house we never thought that this would happen.'

▶ Study the newspaper article above and then write down your answers to these questions.

a Why do you think the Robinsons bought a house so close to the cliff edge?

b What do you think caused the land-slide?

c Is there any way it could have been prevented?

d What do you think has happened to all the bits of rock that have been eroded from the cliff?

▶ Look at these 3 photographs. For each one, write down what you think is causing the erosion of the rocks.

e

f

g

The pieces of rock are often found a long way from where they started. Rock pieces are **transported** by rivers. As they move along, the pieces get smaller and smoother. The further they travel, the more rounded they become.

How are rocks eroded by water?

Make a 'stream' flowing into a 'lake'.
Investigate how the stream moves rocks.

Is the movement affected by:

- how fast the stream flows?
- the width of the stream?
- the type of rock?
- the size of rock?

Make a prediction which you could test.

Write a plan to test your prediction.

How can you make your results reliable?

Show your plan to your teacher and then try it out.

Write a report saying what you did and what you found out.

Was your plan a good one?

How could you improve your investigation?

⚠ if water gets spilt on the floor, mop it up quickly to prevent slipping

Solve the mystery

▶ Look at these 2 photographs. Write down your answers to these questions.

h How do you think erosion has formed the **arch** and the **stacks?**

i Where has the material that was eroded ended up?

1 Copy and complete, choosing the correct word from the 2 given in brackets in each case:
Rocks crumble due to (weathering/erosion) and are then worn away by (weathering/erosion). (Winds/waves) break off pieces of rock when they smash against (hills/cliffs). Glaciers are rivers of (ice/water) that scrape rock out of (mountains/valleys). (Larger/smaller) pieces of rock are carried further away than (larger/smaller) pieces.

2 Design a model to show how waves erode cliffs.
What will you make the cliffs out of?
How will you make the waves hit the cliffs?

3 In very dry countries, winds can pick up sand and blow it against large rocks.
Look at this photograph of a **rock pedestal**.
Try to explain how these are formed in the desert.

4 Ask your teacher for some pebbles from a beach. What sort of shapes do they have? How do you think they have become shaped like this?

Things to do

Settling down

After rocks are broken down, the smaller pieces may be carried away. We say that they are **transported** to another place.

▶ In what ways can the pieces be transported to another place? Look back at page 88 for some ideas.

Eventually the pieces of rock are **deposited** in another area. These pieces of rock are called **sediments**.
The larger sediments are not carried as far in the river. Sediments of a similar size are usually deposited in the same place. Very small pieces of rock can be carried along even when the river flows only slowly. The fine sediment often collects near the mouth of a river (where it meets the sea).
Sediments deposited by the sea may form sand banks.
When a river deposits sediments, they may eventually form a soil.

Near a sandy bay

Where does the soil come from?

▶ Your teacher will give you samples of 4 different soils.
These will be labelled A, B, C and D.
Each of these soils has come from a different place.

Look very carefully at each soil using a hand lens.
Try to match each soil with one of the places shown in these photographs.
Write down the reasons for your choice.

Moorland

Woodland

Farmland

It's a fact!

Sandy soils have large particles and clay soils have small particles. **Loam soils** are a mixture of sandy soil and clay soil. They are easy to dig and hold water without becoming water-logged.
Dead plants and animals decay in soil to form a soft black substance called **humus**.

Investigating soils

Your teacher will give you 2 different soil samples.
You can choose one of these investigations:

A Which soil contains more water?

C Which soil will hold the most water?

B Which soil contains more humus?
(Hint: humus burns off at 110°C.)

D In which soil do seeds grow better?

Plan your investigation. Make sure it is a fair and safe test.

- What equipment will you need?
- How will you record your results?

Looking at sediments

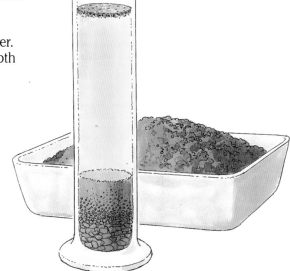

▶ Look very carefully at your rock and soil sample with a hand lens. Are all the particles the same size?

▶ Pour some of your material into a jam jar or measuring cylinder. There should be enough to fill the measuring cylinder to a depth of about 4 cm.

Now almost fill the container with water. Put your hand over the top and carefully shake the container, so that you mix the solid up with the water.

Leave the solid to settle.
Look at it regularly during the lesson.

Write down what you can see.

Have another look at it next lesson.

What does this experiment tell you about sediments?

---

1 Copy and complete:
Soils are made of small . . . . particles. These have broken away from large rocks by . . . . and . . . . Then they may have been . . . . by the action of rivers and streams.
Sediments are small . . . . particles. Some may have been . . . . by the sea to form sand.

2 Try to explain each of the following statements:
a) Sandy soils are easy to dig but need plenty of rain.
b) Clay soils can get water-logged and are then hard to dig.

3 Part of the rock cycle:

Rocks can be broken down by weathering and worn away by erosion. They can then be transported and deposited somewhere else. When they build up as sediments they become squashed and water gets squeezed out. Slowly, a new rock forms.

Make a poster or patterned note of this part of the rock cycle.

4 Estuaries are places where rivers meet the sea. The sediment carried by the river is deposited as mud. Find out about estuaries and mud-flats.

Things to do

Rocks over time

Learn about:
- how layers of sediment form
- evidence from fossils

▶ Read again about igneous, sedimentary and metamorphic rocks on page 85.
Now look back to page 91.
How do you think sedimentary rocks form?

Sedimentary rocks

You know that rocks are weathered.
The small pieces of rock are carried to another place.
They then **deposit** as **sediment**.

Over time layers of sediment pile up.
This puts pressure on older layers underneath.
The pressure pushes the layers of sediment together.
Water is squeezed out of the sediments.
Minerals dissolved in the water are left behind.
They act as a kind of cement between the sediments.
A **sedimentary** rock slowly forms.

Look at the diagram opposite:
There are sharp boundaries between the layers.
This shows that there were time intervals between the
build up of each sediment. Some layers take millions
of years to form, for example, chalk.

At certain times the sea or lake may run dry.

a How could this happen?

This could cause another sediment. Another layer could form.
Rocks formed like this are called **evaporites**.
Try the experiment to see how they form.

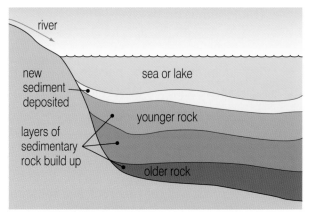

Layers of sedimentary rock forming

This lake has run dry.
A sediment layer is forming.

Evaporating sea water

Half fill an evaporating basin with sea water.
Put the basin on a water bath.
Heat the water bath gently with a medium Bunsen burner flame.
As the water in the basin disappears, what do you see?

Do not evaporate all the water from the basin.
When there is a little water left, switch off the Bunsen burner.
Let the last bit of water evaporate slowly.

b What do you see?

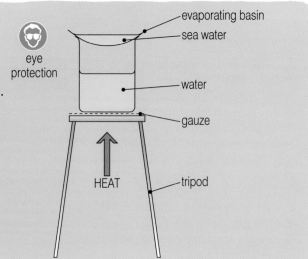

Looking at layers of rock tells us about the climate of the Earth millions of years ago.

Rocks that tell a story

Rocks give us evidence that the Earth is very old ... about 4600 million years old!

Fossils tell us about the animals and plants that lived millions of years ago. They are the remains of animals and plants that have been preserved in rocks.

Some sedimentary rocks are actually made mainly from the remains of plants and animals. Examples include coal (made from trees and ferns that grew in ancient swamps) and chalk (made from the hard bits of ancient sea plants called coccoliths).

▶ Look at the fossil your group has been given.

How do you think this fossil formed?

Use books or ROMs to find out about your fossil.
- When did it live?
- Where did it live?
- What was its environment like?

Write a paragraph about your fossil.

Draw a picture of your fossil and colour it in.

▶ Look at the time chart below. This chart shows 4600 million years of the Earth's history.

Where does your **fossil** fit on the time chart?

Where are **you** on the time chart?

c   Are fossils young or old compared to you?

d   Are fossils young or old compared to the Earth?

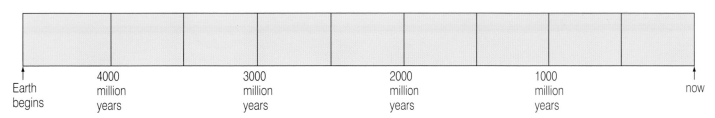

Earth begins   4000 million years   3000 million years   2000 million years   1000 million years   now

---

Things to do

1   Rocks can be broken down into smaller pieces. This is called . . . . The small pieces are . . . . to another place. They can wear away more rock. This is called . . . . The rocks deposit. Layers of . . . . build up. The . . . . pushes the sediment layers together. Water is squeezed out. The . . . . dissolved in the water get left behind. They act as a . . . . between the sediments. A . . . . rock forms. When water . . . . from seas and . . . ., this also leaves sediments which form layers in the rock.

2   Look at this rock sequence:

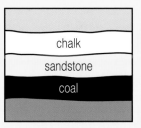

chalk

sandstone

coal

Suggest the order of events that might have produced this rock sequence.

93

Questions

1 Where can you find rock samples?
In what sort of places?
Collecting rock samples can be dangerous.

Imagine you are the teacher of a class.
The class is going to search for rock samples.
You want to make sure the rocks will be collected safely.
Write down some instructions for your class.
Make sure you give safety guidelines.

2 Use reference books or ROMs to find out how coal forms.
Write a paragraph and draw diagrams to explain this process.

3 What do you think these diagrams show?
Copy them out. Try to explain this process.

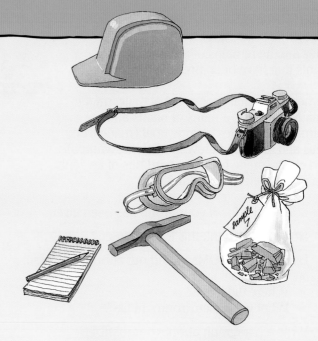

4 Coves form when the sea erodes rocks.
Which is the harder rock around the cove?
Limestone or clay?
Explain how you know this.

5 Sedimentary rocks take a long time to form. The sediments settle in water.
Design an investigation to see how quickly different sediments settle.
Sediments to test could be:
soil, sand, pebbles, silt.

6 Wackham Wanderers are fed up with their soccer pitch.
At the start of the season it's bone dry and full of cracks.
For most of the rest of the season it's water-logged.
a) Plan an investigation to find out how well the water in the soil drains away.
   You can use the sort of equipment found in your science laboratory. Remember to make it a fair test.
b) Suggest ways in which the Wanderers could improve the drainage of their pitch.

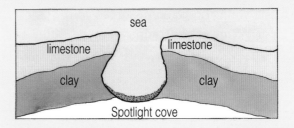

The rock cycle

Have you ever seen a volcano?
When it erupts, red hot lava pours out of the cone.
When the lava cools it forms rock.
But this is only one way of forming rock.
In this unit you will find out about other ways and how
rocks are recycled over time.

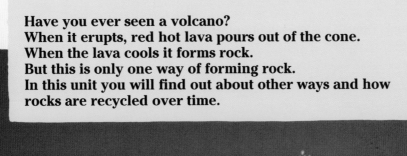

Sedimentary rocks

Rocks are slowly weathered. They break into pieces.

▶ What causes weathering? Draw a spider diagram of your ideas.

Weathering

The weathered rocks are moved by wind, rivers and the sea.
They are **transported** to another place. When these settle (**deposit**),
they are called **sediments**.
The sediments can be fine grains, like sand. They can be larger
fragments, like pebbles.
Over time, layers of sediment build up. The sediment is squeezed
by the weight of new layers above. Any water is squeezed out.
The minerals dissolved in the water are left behind.
These minerals **cement** the grains together.
The solid sedimentary rock forms.

Look at the diagram:

a Which is the newest sediment deposited?

b Which is the oldest rock?

When seas or lakes run dry, sediments form. The water evaporates.

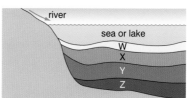

Layers of sedimentary rock forming

Sand and sandstone

Look closely at the samples of damp sand and sandstone.
Use a hand lens or microscope.
How are the grains held together?

c Why are most sandstones porous?

Squash the wet sand. Look for the water being squeezed out.
You can see how the sediments stick together.

Looking at sedimentary rocks

Look carefully at the rock samples your teacher gives you.
These are all sedimentary rocks.
Use books or ROMs to identify them.

Make a list of *features* of sedimentary rocks
(appearance, texture, etc.).

d Name the rock made from pebbles.

e Name the rock made from mud.

f Fossils are often found in sedimentary rock. Why?

Testing limestones

Your teacher will show you some samples of limestones.
Are the limestones the same or different?
Plan the safe tests you will do to answer the questions below.
Be sure you know which measurements to make.

- What do the limestones look like?
- Do they soak up water?
- What are their densities?
- How much carbonate do they contain?

Ask your teacher to check your plans. Then carry out the tests.
Record your results in a table.

g Are all limestones the same?

h Why do you think this is?

i Are your results reliable? How can you improve these tests?

Do some research to find out how two different types of
limestone were formed. Relate your findings to the amount
of calcium carbonate they contain.

▶ Look at the statements below.
They describe how sedimentary rocks form.
Put them in the right order. Copy out your answer.

Rocks can crumble. They make small particles.

The pressure and the mineral cement stick the particles together.

The particles settle in another place. They form a layer.

Water is squeezed out. The minerals left in the sediment act like cement.

The lower layers get pressed together.

The particles are carried away by rivers or wind.

Other layers get put on top of them.

The solid rock forms slowly.

Things to do

1 Draw a cartoon strip to show the process of forming a sedimentary rock. Start with a piece of weathered rock. Finish with the solid lump!

2 A river carries rock fragments from a mountain to the sea. Describe 2 ways in which the fragments change as they are moved.

3 Why can the presence of fossils be useful in the investigation of rocks?

4 Imagine that a limestone quarry is opening near you. You are a local newspaper reporter. Write a balanced report about the quarry. Explain why the quarry is important. Write about problems that might arise.

Mary Anning

Fossils

A **fossil** is any part of something that once lived, and is now preserved in rock.
Only rarely is the fossil a whole body or plant.
Sometimes it may be just a single bone, or even a footprint.

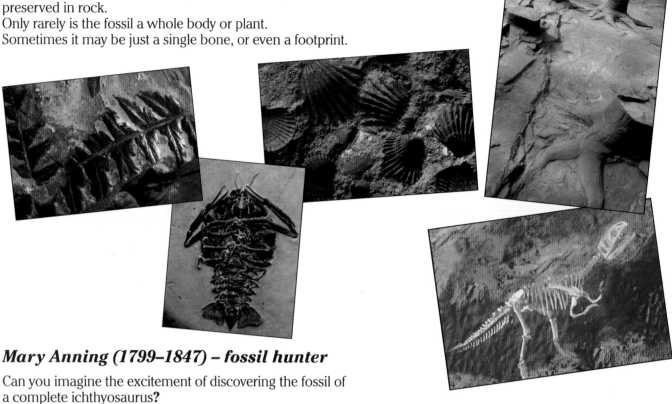

Mary Anning (1799–1847) – fossil hunter

Can you imagine the excitement of discovering the fossil of a complete ichthyosaurus?
Mary Anning had that thrill when she was only 11 years old.

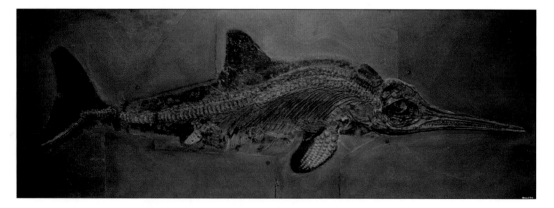

She carefully uncovered a fossil her brother had spotted earlier.
To her surprise the remains of the reptile were 30 metres long!
The fossil is now in the Natural History Museum in London.

Mary has been described as the greatest fossil collector ever.
She was one of the first ***palaeontologists*** (people who study fossils and evidence from prehistoric times).

Mary's father made furniture for a living but the family struggled to make ends meet. It is thought that her mother had as many as 10 children, but only Mary and her brother survived. Her father collected fossils to sell to tourists in their home town of Lyme Regis in Dorset. This helped to bring in a little more money but the family were in debt when he died in 1810.

However, the family were skilled fossil hunters and Mary eventually took over the business.
Even the dog was involved! Mary trained her dog to stay at the spot where she found a fossil while she went off to get her digging tools. Unfortunately the dog was killed in the line of duty when it was buried in a landslide.

Mary did manage to scrape a meagre living for the family with the support of a small grant from the government.

But Mary has not really gained the recognition that her discoveries and great knowledge about fossils deserve. Can you think why?

At that time, men dominated British society. They had all the best jobs and women were treated as second-class citizens.
Poor people were also looked down upon by those with wealth and power. The world of science reflected these attitudes. So you can see why Mary might have struggled to get credit for her work. She was not rich, had received little formal education and was a woman!

However, she did win the respect of scientists of her time. It helped when she discovered the first ever fossil of a plesiosaur. A famous French scientist doubted her discovery. But once he had checked it out himself he agreed that Mary's new fossil was an important find. This helped to get her and her family accepted by fellow scientists as part of their 'club'. She could argue on equal terms with eminent geologists.

She was made an honorary member of the Geological Society of London before she died.

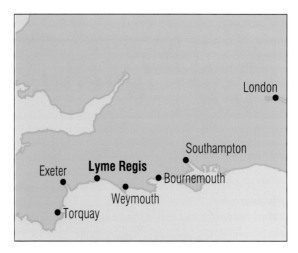

In Victorian times the world of science was dominated by men

It is thought that the tongue-twister: 'She sells sea shells on the sea shore' refers to Mary Anning

---

1 Not much was written about Mary Anning's contribution to science for many years after her death.
Why do you think she has only recently gained recognition for her work?

2 How are fossils formed?

3 Find out why Lyme Regis is a good place to find fossils.

4 Look back at 8G1. Explain the chances of finding fossils in:
a) sedimentary rocks,
b) metamorphic rocks, and
c) igneous rocks.

5 What evidence can we find about the history of the Earth from fossils?
In your answer include information about the theory of plate tectonics.

Things to do

Igneous rocks

Learn about:
- forming igneous rocks
- comparing igneous rocks

Igneous rocks are made from **magma**. This is a molten (melted) material from deep underground.

Magma rises to the surface of the Earth. It can erupt at the surface to form a **volcano**.

The molten rock that erupts is called **lava**. It cools *quickly* at the surface to form solid rock. Some magma never reaches the surface. It cools *slowly* surrounded by rocks underground. It forms solid rock.

▶ Volcanoes can be dangerous.
But some people choose to live on the slopes of volcanoes.
Why do they do this?
Make a list of your ideas.

Igneous rocks are made up of crystals.
The crystals may be tiny or quite large.
What might affect the size of crystal?
Does it depend on how fast the magma cools?
Try the next experiment to find out.

Crystal clues

We can't easily melt rocks in the lab!
Use salol to be your melted rock for this test.

- Half fill a beaker with water.

- Put a solid called *salol* into a test-tube.
  It should be about 3 cm deep.

- Put your test-tube into the beaker of water.

- Heat the water to melt the salol.

- When the salol has melted, switch off your Bunsen burner.

- Ask your teacher for a *cold* glass slide.
  This has been kept in the fridge.

- Use a pipette to put 3 drops of melted salol on to the cold slide.
  Use a magnifying glass to watch carefully.
  You will see salol crystals form.

⚠ Care – the water will still be hot

eye protection

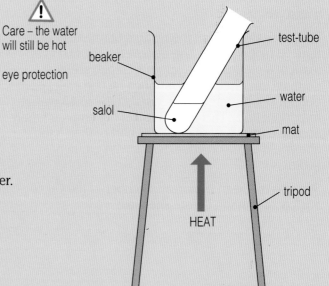

Crystals can form when a liquid cools.

- Now get a *warm* slide. Repeat your experiment.

a On which slide did the salol cool faster?

b On which slide did the bigger crystals form?

c Write a summary of your findings. Include an explanation that uses the particle model to suggest why crystal size differs.

Comparing igneous rocks

Your teacher will give you some rock samples.
Look carefully at the rocks.

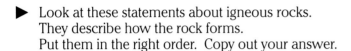

Make a table to show your answers to the following:

- Describe the appearance of each rock.
- Decide which of the rocks are igneous rocks.

For the igneous rocks:

- How quickly did the magma cool to form each rock?
  How do you know?
- Did the rock form by cooling at (or near) the surface, or below ground?

▶ Look at these statements about igneous rocks.
They describe how the rock forms.
Put them in the right order. Copy out your answer.

| Inside the Earth is magma. This is very hot molten rock. |
| The volcano erupts. |
| It cools slowly underground. |
| Lava cools quickly. It makes igneous rocks with small crystals. |
| It makes igneous rocks with large crystals. |
| Then magma comes to the surface of the Earth. It is called lava. |
| Some magma does not get to the surface. |

d Igneous rocks do not contain fossils. Why not?

Igneous rocks are hard. They are used to surface roads. They are usually coated with tar.

Density of igneous rocks

Measure the density of the samples of granite and gabbro.
Granite is 'silica-rich'. Gabbro is 'iron-rich'.

e Which of these rocks is the denser?

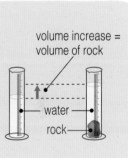

volume increase = volume of rock
water
rock

Now measure the density of the other igneous rocks.
Decide whether the rocks you test are relatively
'silica-rich' or 'iron-rich'.
Remember: density = mass ÷ volume

---

1 Copy and complete:
a) Molten rock is called . . . .
b) Molten rock erupting from a volcano is called . . . .
c) Fast cooling of a liquid makes . . . . crystals.
d) Slow cooling of a liquid makes . . . . crystals.

2 Use books and ICT to help you with this.
a) What is a volcano?
b) Imagine that you are a newspaper reporter. You have been sent to report on an erupting volcano. You are the first at the scene. Write the report for your newspaper.

3 Find out about:
a) volcanoes in the world which are 'active'.
b) a famous volcanic eruption. (How did this affect local people? What happened to the environment?)

Things to do

Metamorphic rocks

Learn about:
● how metamorphic rock forms
● the rock cycle

Metamorphic rocks can be formed from igneous, sedimentary or other metamorphic rocks.

Some rocks are buried deep underground. They are hot. They feel pressure from rocks above them. However, the pressure is increased greatly during 'mountain-building' episodes. During these periods, the Earth's crust can buckle as massive plates of rock collide against each other. Sometimes the heat and pressure are so great that the mineral structures of the rocks change. The particles in the buried rock form new crystals. **Metamorphic rock** is formed.
These changes happen when the rock is still solid. It does not melt.

Sometimes rocks are 'baked' by magma.

a What is magma?

The rocks close to the magma get very hot. The high temperature can change the mineral structure of the rock. It becomes **metamorphic** rock.

The minerals in metamorphic rocks line up in bands

What's changed?

Sedimentary rocks can become new metamorphic ones.
Your teacher will give you some rock samples.
Look at the rocks with a hand lens.
Match the **sedimentary** rock with the **metamorphic** rock it makes.

Choose one matching pair of rocks.
Describe the differences between the rocks.

▶ Look at this rock face diagram:
It shows 3 types of rock.

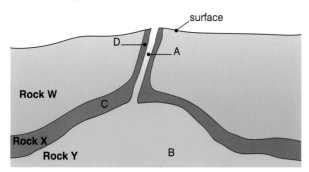

b Which rock is sedimentary?

c Which rock is igneous?

d Which rock is metamorphic?

e The rock at A cooled more quickly than the rock at B. How does that affect its appearance?

f Why is the band of rock thicker at C than at D?

3 rock types!

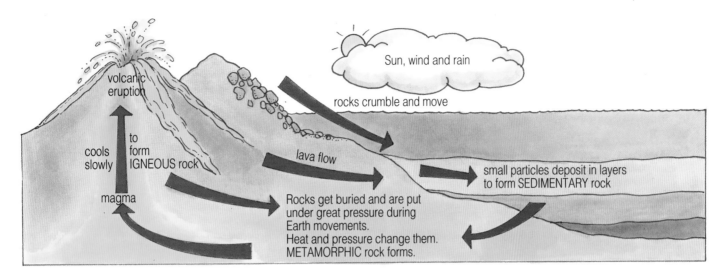

volcanic eruption

cools slowly — to form IGNEOUS rock

magma

lava flow

Rocks get buried and are put under great pressure during Earth movements.
Heat and pressure change them.
METAMORPHIC rock forms.

Sun, wind and rain

rocks crumble and move

small particles deposit in layers to form SEDIMENTARY rock

g This is called the **rock cycle**. Why do you think this is**?**

h Draw a flow diagram to represent the rock cycle.

Identifying rocks

These could be features to look for:

Igneous	Sedimentary	Metamorphic
Hard	Can be soft and crumbly (but not always!)	Hard
Made of crystals No layers	Made of lots of grains	Often made of bands or sheets Splits into layers
No fossils	May contain fossils	May contain distorted fossils

- Use these ideas to sort some rock samples into groups.
- Name as many of the rocks as you can.
- Use books or ROMs to list some uses of these rocks.

1 Copy out the 2 lists of rocks.
Use a line to join each sedimentary rock with the metamorphic rock it makes.

Sedimentary	Metamorphic
limestone	quartzite
sandstone	marble
shale	slate

2 Fossils found in metamorphic rock are often distorted. Why?

3 You must get permission for this first!
Visit a local garden centre or builder's yard.
Make a list of the rocks you see.
How are the rocks used?

4 Find out some information about ***earthquakes***.
a) What causes earthquakes?
b) What is the ***epicentre*** of an earthquake?
c) What is the ***Richter scale***?
d) What is a ***seismometer***?
e) Why is San Francisco at risk from earthquakes?

Things to do

8H5

Chemistry at Work

Limestone

Limestones are sedimentary rocks. Some are made from the shells of sea creatures

a   Describe how shelly limestone was formed.

Limestone contains calcium carbonate.
Calcium carbonate has the chemical formula $CaCO_3$.

b   How many elements make up calcium carbonate?

Limestone is a very important rock.
Blocks of it are used to construct buildings.

c   Why are many old limestone buildings in cities
showing bad signs of weathering?

Limestone cottages

Limestone is used to make cement, mortar and concrete for these buildings

We also use limestone to make other building materials,
such as *cement*. Cement is then mixed with more rock
products to make *mortar* and *concrete*.

When we make *cement* the limestone is ground up into
a powder and heated with clay. A little calcium sulphate
is then added to it.

d   Find out how clay is formed in nature.

The mixture is heated in large rotating kilns.
Look at the diagram opposite:

When limestone is heated it breaks down into
calcium oxide and carbon dioxide gas is given off.

e   Write a word equation to show what happens
when limestone is heated.

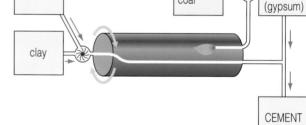

Mortar is made by mixing cement powder with sand,
then adding water to make a thick paste.
It is the material that binds the bricks together when
a house is built.

Concrete is made from a mixture of cement, sand
and gravel or small stones. As with mortar, we then
add water and mix it all together.
Reinforced concrete is made when we need large blocks,
such as for building motorway bridges. The concrete
is set in moulds with iron rods running through them.

f   What might happen if the iron rods inside the
concrete blocks started to rust?

Using mortar

Using concrete

Minerals

A mineral is a substance found in the Earth's crust.
Rocks are usually mixtures of minerals.

Mohs' scale is used to compare the hardness of different minerals.
It was first compiled by Fredrich Mohs, a German scientist,
in 1822.

Look at the table below:

Talc is very soft

Number on Mohs' scale	Mineral	Approximate relative hardness
1	Talc	1
2	Gypsum	3
3	Calcite	9
4	Fluorite	21
5	Apatite	45
6	Orthoclase, feldspar	72
7	Quartz	100
8	Topaz	200
9	Corundum (ruby and sapphire)	400
10	Diamond	1600

Rubies are hard minerals

g How many times harder than ruby is diamond?

Look at the photos below:

Engraving glass

Diamond tipped mining cutters

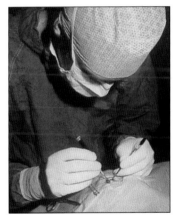

A surgeon using diamond coated instruments

h Why is diamond chosen for these uses?

Gemstones are minerals that are hard and attractive.
They can be cut along certain planes to make flat surfaces on the stone.
Diamond is a very expensive gemstone.

i Quartz is a mineral found in sand.
Why are most gemstones over 7 on Mohs' scale?

j Do some research to find out about one particular gemstone.
You might choose your birth-stone.
Try to find what it is made from, how hard it is and where
it is found.

Questions

1 Write a worksheet for Year 8 pupils.
Show how to test rocks for:

- hardness
- porosity (how easily they soak up water)
- density.

You should include diagrams and instructions.
You could use cartoons. You could include questions too.

2 A gem is a precious mineral. It can be cut and polished.
Then it can be used in jewellery.
   a) What makes a mineral precious?
   b) Name as many gems as you can.

3 Explain how each of these rocks is formed:
   a) granite
   b) conglomerate
   c) slate.

4 Sea defences can be used to limit coastal erosion.
Concrete can be used to build them.
Concrete is made from sand, cement and gravel mixed with water.
Plan an investigation to test different concrete mixes.
Which will make the best sea defence?
(You must be able to do your tests in the lab.)

5 Look at the simple rock cycle:
Choose words from the list below to put labels in the boxes.

heat   melting   erosion   deposition   pressure
weathering   transportation   crystallising

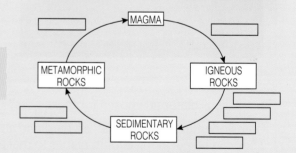

6 Find out about geological time periods.
During your research, try to answer the questions below.
   a) In which period were there lots of sea animals, but no life on land?
   b) In which period did the Ice Age occur?
   c) Which period lasted from about 135 million to 65 million years ago?
   d) In which period did reptiles start to appear on Earth?
   e) Which geological time period are we in today?

In the cretaceous period, great reptiles roamed the Earth

Heating and cooling

Some people confuse the words 'temperature' and 'heat'. In this unit you will find out the difference.

You will also find out about the transfer of heat, by conduction, convection and radiation.

And you will investigate what happens when substances are heated and cooled.

Warming up

Learn about:
- the Celsius scale
- temperature and heat
- how heat is transferred

a liquid-in-glass thermometer (with a scale in degrees Celsius)

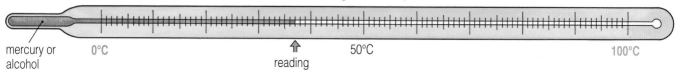

mercury or alcohol   0°C   ⬆ reading   50°C   100°C

a   What do you measure with a **thermometer?**

b   What is the reading on this thermometer**?**
(This is the temperature of a human body.)

c   What happens to water at i) 0°C**?**   ii) 100°C**?**

d   Explain how you think this thermometer works.

> **What's the temperature?**
> Look at some different thermometers.
> - Discuss how you think each one works.
> - Use them to measure some temperatures.
>
> - Carry out a survey to find out how accurately people can estimate temperatures.
> - Analyse your results, taking into account the size of your sample.

Heat (energy) and temperature

Thermal energy (heat) is **not** the same thing as temperature.
To understand this, let's compare these 2 things:

A white-hot spark

The tiny sparks are at a very high temperature, but contain little energy because they are very small.

Each particle in the spark is **vibrating**.
Because it is very hot they are vibrating a lot.
But there are not many of them, so the total amount of energy is small.

This idea about particles is called the **Kinetic Theory**.

A bath-full of warm water

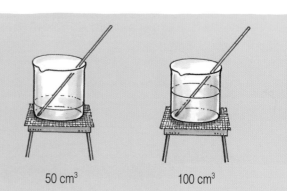

The water is at a lower temperature, but it contains more energy.
This is because it contains more **particles**.

Each particle is vibrating at a low temperature, but there are many of them.
There is a lot of thermal energy (heat).

Warming up water

Plan a safe investigation to see what happens when you give the same amount of energy to different amounts of water.

- Make a prediction.
- How will you make sure that you give the **same** amount of energy each time**?**
- Show your plan to your teacher, and then do it.
- Explain what happens, using these words:
  energy   particles   vibrating   temperature

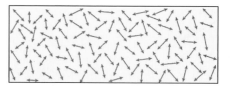

50 cm³   100 cm³

Energy on the move

Energy always travels **from** hot things **to** cold things.
There are 3 ways that the energy can be transferred:

Conduction

The metal handle gets too hot to hold.
The energy has been **conducted**
through the metal.

The metal is a good **conductor**.
The wooden spoon is an **insulator**.

Convection

The air over the heater is warm.
The hot air is rising upwards, in
convection currents.

You get similar convection currents
when you heat a beaker of water.

Radiation

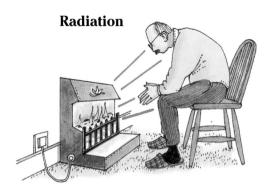

Radiant energy is travelling through the
air, just like solar energy from the Sun.

The rays can travel through space, at
the speed of light. They are also called
infra-red rays.

▶ Here is a picture of a Bunsen burner heating up some gauze:

e After a while, the base feels warm at the point A. Why is this**?**

f If you put your hand at point B, it is hot. Why is this**?**

g If you put your hand near the red-hot gauze, at point C,
it feels warm. Why is this**?**

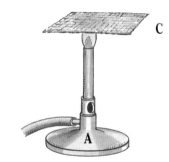

▶ Here is a 'model' to help us see the difference between these 3 ways.
Three ways of getting a book to the back of the class:

1 Conduction: you can pass a book
from person to person – just as the
energy is passed from atom to
atom.

2 Convection: you can carry the book
to the back of the class – just as hot air
moves in convection, taking the
energy with it.

3 Radiation: a book can be thrown
to the back of the class – rather like
the way energy is radiated from a
hot object.

---

1 Copy and complete:
a) When an object is heated, the energy
makes the . . . . vibrate more. The hotter
the object, the more the . . . . vibrate.
b) Energy travels from hot objects to . . . .
objects by . . . . , or . . . . , or . . . .

2 Explain why a white-hot spark falling
into a bath of water does not make the
water hot.

3 Explain how an electric fire heats up a
room.

Things to do

Conduction

▶ What would you feel if you stirred some hot soup with a metal spoon and with a wooden spoon? Can you explain this?

Energy on the move

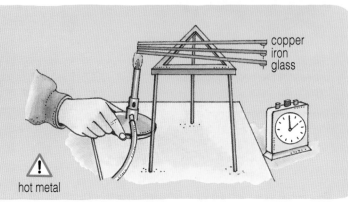

copper
iron
glass

● Set up 3 rods as shown, of copper, iron and glass:

● Use vaseline to fix a drawing-pin at the end of each rod.

● Then heat the ends of the rods equally, with a Bunsen burner.

● What happens? How long does it take for the first and second pins to fall off?

⚠ hot metal

a Which way was energy flowing in the rods?

b Does the energy flow at the same rate through all the rods? Explain your answer.

c Which of these 3 materials would be best for making a pan? Explain your answer.

▶ The diagram shows how we can explain conduction, using the idea of particles:

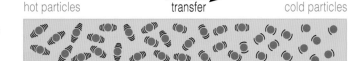

high temperature
hot particles

heat
transfer

low temperature
cold particles

Energy is transferred from particle to particle along the bar. At the hot end, the particles are vibrating a lot. As they bump into each other, the energy is passed along the bar.

▶ As you saw, copper is a good **thermal conductor**. In fact, **all metals are good conductors**. Glass is a poor conductor. It is an **insulator**. Water is a poor conductor. Air is a very good insulator. We use this to keep us warm:

Birds fluff up their feathers in winter, to trap more air. The air is a good insulator and keeps them warm.

An anorak has a lot of trapped air. This slows down the transfer of thermal energy from your body.

Insulation in the loft of your house keeps you warm, and saves money. The material contains a lot of air.

Keeping warm

Your teacher will give you some materials that could be used for clothes, or to insulate your house.

Your job is to find out which of these materials is the best insulator.

hot water

- You could use a thermometer, or you could use a temperature sensor connected to a computer. What other equipment will you use?

- How will you make it a fair and safe test?

- How will you record your results?

- Show your plan to your teacher, and then do it.

- Which material is best? **Why** do you think it is best?

temperature sensor

Saving energy at home

Here are 2 houses:

One of them has been carefully insulated.

d Which house has **not** been insulated?

e What is the total heating bill for this house?

f Imagine you lived in this house. Which parts would you insulate first?

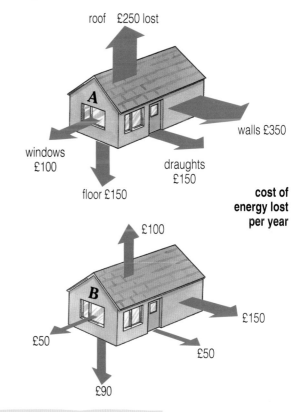

roof £250 lost

walls £350

windows £100

draughts £150

floor £150

cost of energy lost per year

£100

£150

£50

£50

£90

g What is the total heating bill for the insulated house?

h All these places have been insulated: *walls, roof, floor, doors, windows.*
- Put them in order of the money saved.
- Next to each one, write the amount of money saved.

i Draw an Energy Transfer Diagram for each house.

---

1 Copy and complete:
a) All metals are good . . . .
Copper is a very good . . . .
b) Glass is an . . . .
Air is a very good . . . .
c) Thermal energy passes through the bottom of a pan by . . . .
The energy is passed from each vibrating particle to the next . . . .

2 Make lists of where a) conductors and b) insulators are used in your home.

3 Write a letter to a pupil in your last school explaining what conduction is.

4 A double-glazed window has 2 sheets of glass, with air between.
Plan an investigation to see if double-glazing helps to insulate.
(Hint: you could use a beaker inside a bigger beaker.)

5 Explain this statement: "All the energy used in heating our homes is wasted."

Things to do

Convection and radiation

Convection

Have you ever noticed that flames always go upwards?
This is because hot air is less dense than colder air.
The hot air rises.

a Where is the hottest part of the room – the floor or
the ceiling? Why?

b Why does smoke go up a chimney?

hot air rises

hot gas expands, takes up more space, so it is less dense, and so it rises.

cold gas particles are closer together

- Fill a beaker with cold water.
- Using tweezers, very gently, place a crystal of purple dye (harmful) at the bottom and near the side:
- Put a **small** flame under the crystal.
- What happens? Explain what you see.

The water moves in a **convection current**.
This carries the energy round the beaker.

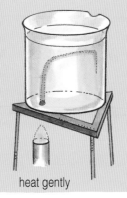

eye protection

heat gently

▶ You get the same thing in a room.
The room is heated by the convection currents
moving round:

c Why does a hot fire sometimes give you a cold
draught on your feet?

On a sunny day, hot air currents can rise from the
ground. Glider pilots can use them to lift their wings.

The Sun can cause very large convection currents,
which we feel as **winds**.

▶ Use what you know about convection currents to
explain what is happening in these photos:

convection current

Radiation

This sun-bather is getting hot:
Her body is ***absorbing*** energy.

d Where is the energy coming from?

e Could the energy have reached her by
conduction or convection? Explain
your answer.

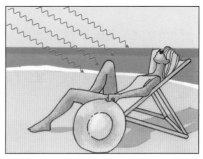

Heat energy can only be transferred through a vacuum (like space) by **radiation**

This energy is called **solar energy** or **radiant energy**.
The rays include **infra-red rays** and **ultra-violet rays**.

f How can you use solar energy to cook food?

▶ Our bodies also ***emit*** (give out) radiation.
We ***radiate*** energy.
This is shown on the thermogram:

g Which part of the man is giving out
  i)   the most energy?
  ii) the least energy?

h Use the key to estimate the temperature of his cheek.

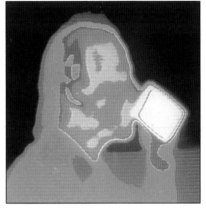

☐	above 38°C
	35°C
	32°C
	29°C
	26°C
■	below 23°C

Melanie and Chris are discussing the colour of cars.

Melanie says, "I think black cars get hotter in the Sun."
Chris says, "Silver is brighter – I think a silver car will
get hotter."

Plan an investigation to see who is right.

• How will you make it a fair and safe test?

• How many readings will you take?

• How will you show your results?

Show your plan to your teacher before you do the investigation.

1 Copy and complete:
a) Thermal energy (heat) can be carried
through a liquid by a . . . . current. The
hot liquid . . . . and the . . . . liquid falls.
b) . . . . currents also flow in air.
c) . . . . rays travel from the Sun through
empty space. This energy is called
. . . . energy or . . . . energy.
d) A black object absorbs more . . . . than
a silver one.
e) A silver surface . . . . the rays like a
mirror. This is used in a . . . . cooker.

2 Explain why:
a) Food cooks faster at the top of an oven.
b) Fire-fighters enter smoke-filled rooms by
crawling.
c) Houses in hot countries are often white.
d) There is shiny metal behind the bar of an
electric fire.

3 A potato is being cooked in boiling water.
Explain, as fully as you can, how energy gets
from the gas flame or hot-plate into the
middle of the potato.

Things to do

Energy for a change

Learn about:
- melting and freezing
- boiling and condensing

solid   liquid   gas

Water is usually a liquid. But water can be a solid (ice) or a gas (steam).
Ice, water and steam are the same chemical substance (H_2O).
When water turns into ice or steam we say it is **changing state**.

▶ Sketch the pictures of the 3 states of water. Next to each arrow,
write a label. Choose the correct label from this list:

> ***boil (or evaporate)   melt   freeze   condense***

a   At what temperature does water boil?
b   At what temperature does ice melt?

distillation
is
evaporation + condensation

Is this true? Explain.

Liquids into solids

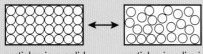

particles in a solid   particles in a liquid

How do you turn a solid into a liquid? How do you turn a liquid into a solid?
In this experiment you will make a solid melt by heating it. Then you
will let the liquid cool slowly. What do you think will happen?

- Put a few spatula measures of the solid in a test-tube.
  Clamp the tube in a warm water bath.

- Heat up the water bath until the solid melts.
  When the temperature of the liquid reaches 75°C,
  switch off your Bunsen.

- Leave the substance to cool. Record the temperature
  of the substance every minute as it cools.

- Stir carefully. Take readings every minute until
  the temperature reaches 50°C.

- Leave the apparatus to cool.

⚠ hot
eye protection

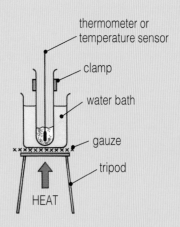

thermometer or
temperature sensor

clamp

water bath

gauze

tripod

HEAT

Plot a graph of your results.
Try to explain the shape of your graph.

What happens during the flat part of
your graph?

temperature (°C)

time (minutes)

Liquids into gases

Dip your finger into water.
Dip the same finger of the other hand into alcohol.
Hold your fingers out in front of you.

drops of water

drops of alcohol

⚠ Alcohol is highly flammable. Any Bunsen MUST be extinguished

c Which liquid cools your skin most?

d Which liquid evaporates faster?

e Try to explain these results using the idea of moving particles.

f What happens to the temperature of a liquid as it evaporates?

g You have thermometers, rubber bands, paper tissues, beakers and a stop-clock.
How can you use this apparatus to find out which liquid evaporates fastest – alcohol, water or propanone?

Feeling cool?

Solids into liquids

Hassan studied some ice melting.
He stirred crushed ice in a beaker.
He took the temperature of the ice every 2 minutes.
These are his results:

thermometer
beaker
ice

Time (minutes)	0	2	4	6	8	10	12	14	16	18	20
Temperature (°C)	−4	−2	0	0	0	0	0	0	0	2	5

h At what temperature did the ice melt?

i Ice melts when it is heated. Where did the heat come from in this experiment?

j Why didn't the temperature rise between 4 and 16 minutes? Explain what happens to the particles.

Salt lowers the freezing point of water.

k Why do we put salt on icy roads in winter?

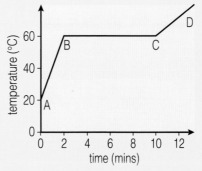

Things to do

1 Copy and complete using **heat** or **cool**:
a) To turn a solid to a liquid you need to . . . . it.
b) To turn a liquid to a gas you need to . . . . it.
c) To turn a gas to a liquid you need to . . . . it.
d) To turn a liquid to a solid you need to . . . . it.

2 Use the words from the box to describe
a) melting point b) boiling point.

solid   liquid   gas   temperature

3 We get rain because water from the sea forms clouds which give us rain.
Use the words below to describe this:

evaporate   condense   Sun   energy

4 Look at this heating curve for a solid substance X:

a) What is the state of substance X from:
i) A to B? ii) B to C? iii) C to D?
b) What is the melting point of X?
c) X is being heated. Why does the temperature stay steady from B to C?
d) What do you think happens if heating carries on after 12 minutes?

Ideas about energy

Learn about:
● early ideas about energy
● how ideas change over time

Ideas about energy started to become important when engineers like **Thomas Newcomen** (in 1705) and **James Watt** (in 1764) tried to improve steam engines. They wanted to know how to make their engines more efficient, so that they wasted less energy.

James Watt's 1785 steam engine

A replica of George Stephenson's '**Locomotion**' of 1825

At that time the scientists didn't know what energy was, but they started to investigate.

In 1760, **Joseph Black** showed that temperature and heat are not the same thing.
He also developed the 'caloric' theory. This theory said that heat was a kind of invisible liquid, that flowed out of hot objects and soaked into cold objects.
At first this seemed a good theory, because it explained most experiments. For example, it explained why mixing water from a hot tap with water from a cold tap gives warm water.

But it couldn't explain everything – it couldn't explain why your hands get warm when you rub them together quickly.

Nevertheless, scientists kept to this 'caloric' theory as the best theory they had.

Joseph Black

In 1798, **Count Rumford** (Benjamin Thompson) was in charge of drilling the holes in some cannons.
While a gun was being drilled, he noticed that it got very hot (by friction).

By drilling for a long time and taking measurements of the temperature, he showed that the gun continued to heat up, long after it should have run out of 'caloric liquid'.

He did more experiments to show that the heat could not have come from the metal or from the air, but only from the movement energy of the drill.
As he wrote: '*It appears to me to be extremely difficult to form any distinct idea of heat, except it be* **motion**.'

Count Rumford

James Joule was born in Salford, Lancashire in 1818. As a teenager he was taught by **John Dalton** (the man who suggested the idea of atoms in 1803).

In 1837, Joule began a series of precise experiments on energy. To do this he had to make his own thermometers which were accurate to $\frac{1}{50}$°C.

In his electrical experiments he looked at the heating effect of a current in a wire.
He found that the heating depends on the *square* of the current. So twice the current gives four times as much heat.

James Joule

In his most famous experiment, he built a pulley system so that falling weights turned some paddles that were in a pan of water. As the weights fell, the turning paddles heated up the water (by friction), and Joule measured the rise in temperature.
He repeated the experiment many times, and measured the temperature very accurately.

From this he predicted that the water at the bottom of a waterfall would be slightly warmer than the water at the top.
Can you see from the labels on the diagram why this should happen?
In 1847 he married Amelia, and they went on honeymoon to the French Alps. Joule spent a lot of his honeymoon measuring the temperatures at the top and bottom of waterfalls.

From all his experiments, Joule got evidence to convince other scientists that energy is 'conserved'.
This is the famous **Law of Conservation of Energy**:
  energy can be changed from one form to another, but it cannot be created or destroyed.
  The total amount of energy stays the same.

This is shown on an Energy Transfer Diagram:

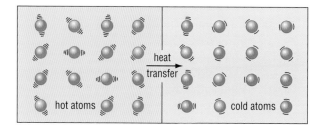

Later it became clear that heat is really the movement energy of atoms and molecules (see page 110).
This is the **Kinetic Theory**.

This idea of moving particles helps us to explain why heat is conducted from a hot object to a cold object, as shown by this diagram:

Conduction from a hot object to a cold object:

hot atoms   heat transfer   cold atoms

---

1 Why were ideas about energy important to people making steam engines?

2 What was the problem with the 'caloric' theory?

3 What happens to the heating effect in a wire if the current increases
a) to 3 times as much?
b) to 10 times as much?

4 From the description above, try to sketch a diagram of Joule's 'paddle-wheel in a pan' apparatus.

5 Use the diagram above to explain why you would expect the water to be slightly warmer at the bottom of a waterfall.

6 Explain how '*heat is really the movement energy of atoms and molecules*'.

Things to do

Questions

1

Look at the picture above, and find as many examples as you can of conduction, convection, radiation. List and explain each one briefly.

2 Design a leaflet for Year 7 pupils that explains heat transfer, using examples they would be interested in finding out about.

3 Jo has 2 mugs. They are the same except that one is black and one is white. They were filled with hot coffee and allowed to cool. Jo took their temperatures every 2 minutes, as shown in the table:

a) Plot a graph (of temperature against time) for each mug, on the same axes. Draw the lines of 'best fit'.
b) Which result do you think is wrong?
c) What is the temperature of the black mug after 3 minutes?
d) What is the difference in temperature after 9 minutes?
e) What conclusion can you draw from the graphs?

Time (min)	Black mug (°C)	White mug (°C)
0	90	90
2	68	78
4	55	67
6	45	58
8	37	52
10	31	43
12	26	37

4 Theo has 2 insulated pans of water at 20°C.
He gets 2 identical blocks of iron, both at 100°C, and puts one in pan A and one in pan B.
He measures the 'final' temperature that the water rises to, and puts his results in the table:

a) Explain carefully why the rise in temperature is less for pan B.
b) What is the final temperature of (i) the iron block in pan A? (ii) the iron block in pan B?
c) The 2 iron blocks were identical. Which one transferred more heat energy to the water? Explain your reasoning.
d) Explain the difference in the iron atoms when they were hot and when they were cold.

	Temperature of the water at the start (°C)	Temperature of the block at the start (°C)	Volume of water (cm³)	'Final' temperature of the water (°C)
A	20	100	500	35
B	20	100	1000	28

Magnets and electromagnets

Magnetism is important to all of us.
Our lives would be very different without it.

We can use electricity to make magnets – which we use every day in radios, TVs, cassette and CD players, phones, and many other things, as you can see below:

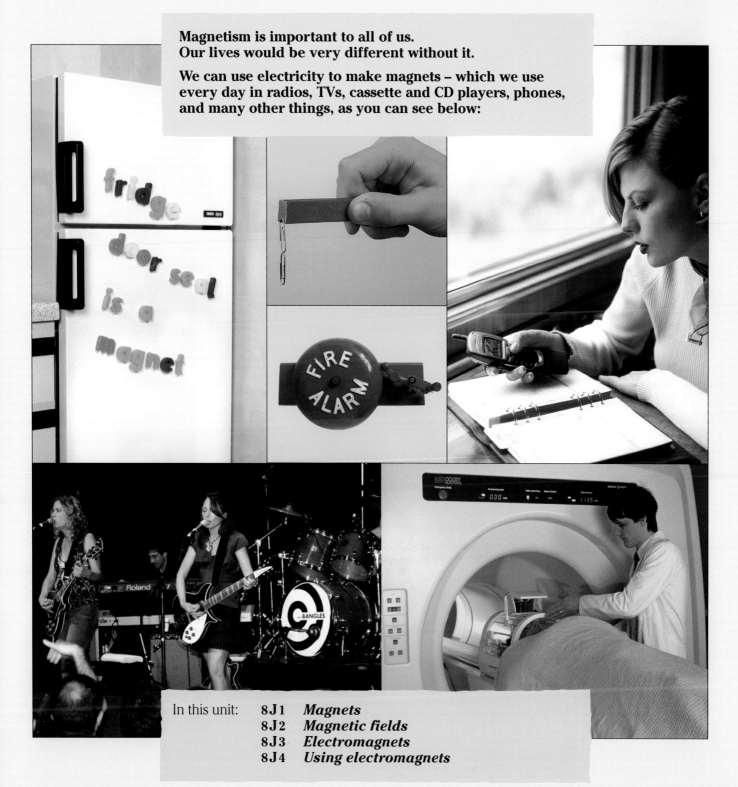

Magnets

a If you spill some pins on the floor, what is the best way to pick them up?

If you use a magnet, the pins stick to the ends or **poles** of the magnet.

b Can you use a magnet to pick up paper off the floor? Why not?

Sailors used to hang a magnet from a piece of string, so that it could swing freely, as shown:

c Why did the sailors do this?
d What is the name of this instrument?

a compass

The end of the magnet that points to the North is called the North-pole (N-pole) of the magnet.

e What is the name given to the other end of the magnet?

Some materials are magnetic, and some are not.

f Which of these are magnetic: wood, iron, plastic, paper, steel, rubber, copper, brass, nickel, cobalt, iron oxide?

g Are all metals magnetic? Which ones are?

Making a magnet

A piece of iron or steel can be magnetised by stroking it several times with a magnet, as shown:

h How could you test the strength of the magnet you make? Try it if you have time. How many times should you test it?

i Why is it sometimes useful to have a screwdriver which is magnetised?

If a magnet is heated until it is red-hot, it becomes **demagnetised**.

Magnetic fields

Magnets can *attract* (pull) or *repel* (push) other magnets.
A **N**-pole repels another **N**-pole.
A **S**-pole attracts a **N**-pole.

They can do this without touching. This is because a magnet has a **magnetic field** round it. Iron and steel are affected by a magnetic field.

The Earth has a magnetic field round it. This field makes a compass point to the North.

j What happens if a **S**-pole is brought near to a **S**-pole?

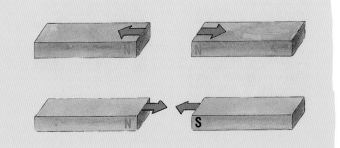

Investigating magnetic games

Here are 4 magnetic games. Plan your time carefully to do as many as possible.
For each one, draw a sketch and write down a *scientific* description of what happens.

Race track

Draw a race track and then 'drive' a paper-clip round it.

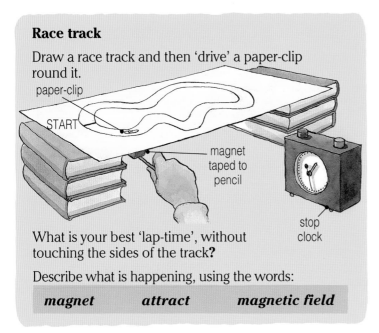

What is your best 'lap-time', without touching the sides of the track?

Describe what is happening, using the words:

magnet   attract   magnetic field

Coin sorter

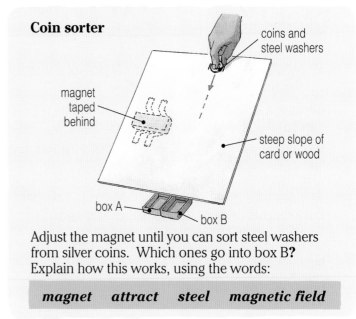

Adjust the magnet until you can sort steel washers from silver coins. Which ones go into box B?
Explain how this works, using the words:

magnet   attract   steel   magnetic field

Identikit face

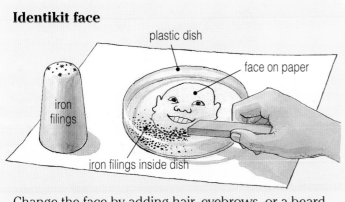

Change the face by adding hair, eyebrows, or a beard.
Describe what you are doing, using the words:

magnet   attraction   iron   magnetic field

Magnetic dogs

Make two 'dogs' like this:

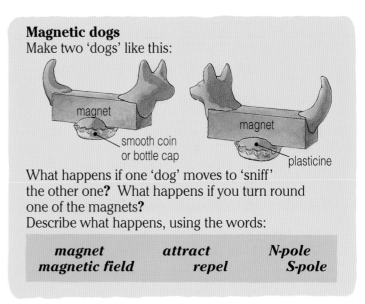

What happens if one 'dog' moves to 'sniff' the other one? What happens if you turn round one of the magnets?
Describe what happens, using the words:

magnet   attract   N-pole
magnetic field   repel   S-pole

Things to do

1 Copy and complete:
a) A magnet has a . . . . field round it.
b) The . . . . field is strongest near the ends of the magnet, called the North-. . . . and the South-. . . .
c) A piece of iron can be magnetised by . . . . it with a . . . . It can be demagnetised by . . . . it.
d) The Earth has a magnetic . . . . round it.
e) A N-pole . . . . another N-pole.
A S-pole . . . . a N-pole.

2 Suppose you are given a bowl of sugar with some iron filings mixed up in it.
a) How could you separate the sugar from the iron filings?
b) Can you think of a completely different way of doing this?

3 Design an investigation to compare the strengths of two bar-magnets.
What equipment would you need? Draw a diagram.
How would you make it a fair test?

Magnetic fields

Learn about:
● plotting magnetic fields
● Earth's magnetic field

a   What happens if you bring the N-pole of a magnet near the N-pole of another magnet?

Because a magnet can repel without touching, we say it has a **magnetic field** round it.

We cannot see the magnetic field round a magnet, but we can find out the shape of it.

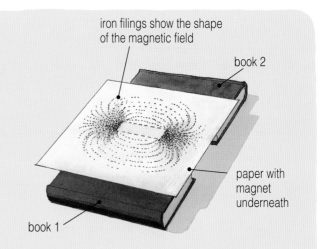

iron filings show the shape of the magnetic field

book 2

paper with magnet underneath

book 1

Put a magnet under a sheet of paper as shown:

Sprinkle some iron filings over the paper and then tap the paper.

Look carefully at the pattern that appears. Can you see that it is the same shape as in the diagram?

Make a sketch of the shape you get.

The iron filings act as tiny compasses, and point along the magnetic field.
The curved lines are called **field lines** or **lines of flux**.

● Where is the field strongest?
  Where are the field lines closest?
● Where is the field weakest?

Here is a better way. Use a ***plotting compass*** to make a map of the magnetic field.

Follow these instructions carefully.

1   Place your magnet on a large sheet of paper and draw round it to mark its position.

2   Choose a starting point near the N-pole of the magnet and mark it with a pencil dot.

3   Put the 'tail' of the compass pointer over your dot, and then draw a second dot at the 'head' of the compass pointer.

4   Move the compass along until its tail is over the dot, and continue in the same way.

5   Dot the path of the compass as it leads you through the magnetic field. Join up the dots to make a smooth line.

6   Choose another starting point, to get different field lines, as in the diagram.

● Is it the same shape as before?

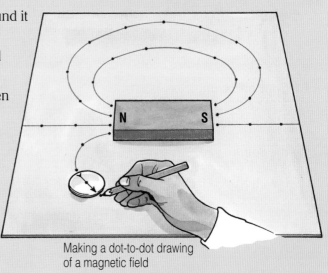

Making a dot-to-dot drawing of a magnetic field

The Earth's magnetic field

b How do you know that the Earth has a magnetic field?

c Which way do the field lines point in this room?

The diagram shows the magnetic field round the Earth:

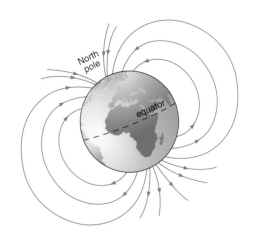

d What do you notice about the shape of the Earth's field?

e Near the North Pole of the Earth, is there a magnetic N-pole or a magnetic S-pole? Why?

f A compass can be made by suspending a magnet from some string (see page 120).
Can you devise a better compass, with less friction?
(Hint: could you use a flat piece of cork?)

Testing magnetic materials

Clamp a magnet so that it levitates a paper-clip,
like an 'Indian rope-trick', as shown:

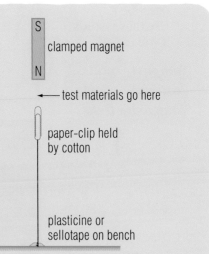

- Do you think it is possible to shield the paper-clip from the magnet's field, so that the paper-clip falls down?

- *Predict* what you think will happen with different materials (e.g. paper, aluminium-foil, wood, plastic, steel sheet, etc.).

- Then try it, with different materials.

- What pattern do you find?

A challenge

Imagine you are given 3 metal bars that look identical.
You are told that two of the bars are magnets, and the other is an unmagnetised bar of steel.

Your task is to work out a method to decide which bar is which.

You should be able to work out *three* different methods to do this.

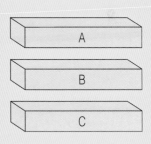

- In your group, discuss which is the *best* method.

---

1 Copy and complete:
a) A magnetic field can be plotted with iron . . . . , or with a plotting . . . .
b) The magnetic field is strongest where the field lines are . . . .
c) A magnetic force cannot act through any material which could be made into a . . . .
d) The best test to see if an object is a magnet is to see if it can . . . . a magnet.

2 Explain why, near the North Pole of the Earth, there must be the S-pole of a magnet.

3 Imagine you are given a small plotting compass and 3 grey metal bars. One bar is aluminium, one is unmagnetised iron, one is a magnet. How can you identify each one?

4 Research the work of William Gilbert, Queen Elizabeth I's doctor.
What did he discover about the Earth?

Things to do

Electromagnets

Learn about:
- making an electromagnet
- exploring magnetic fields

Using magnets

Here are some uses of magnets:

▶ For each one, write down a sentence to describe it.
Use these words if you can:

magnet   pole
attract   repel
magnetic field

a   Cupboard door catch

c   In an eye hospital

b   A compass

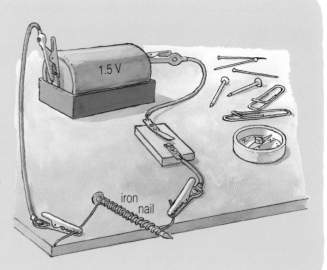

d   Magnetic crane in a scrap-yard

Making an electro-magnet

The door-catch and the compass use permanent magnets.
But the crane uses an **electro-magnet**.
An electromagnet can be switched on and off.

▶ Use the diagram to make your own electromagnet:

Warning: only connect the battery for a few seconds or it will soon go flat!

- How can you test the strength of your electromagnet? Try it.

- What happens when you switch off the current?

Electromagnets are used in all electric motors, door-bells, loudspeakers and TV sets.

1.5 V

iron nail

Electro-magnetism

In 1820, a Danish scientist called Hans Oersted discovered that:
an electric current produces a magnetic field.

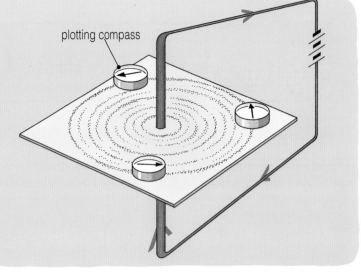

plotting compass

In this experiment, a large current is passed up the thick copper wire.
Iron filings are sprinkled on the card to show the shape of the magnetic field:

Plotting compasses show the direction of the field lines.

e What happens if the current is reversed?

f What happens to the compasses when the current is switched off?

The magnetic field from a single wire is very weak. To make it stronger, the wire is made into a **coil**, or **solenoid**.

The field round a coil

Iron filings are sprinkled on a card round a coil:

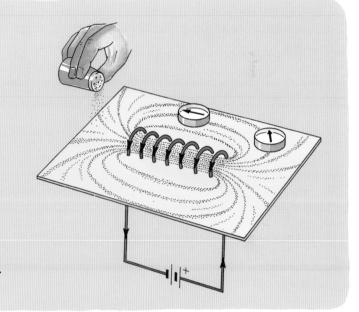

g What happens when the current is switched on?

h What do you notice about the shape of the magnetic field? (Hint: see page 122.)

The compasses show the direction of the magnetic field.

i What happens if the current is reversed?

j What happens to the compasses when the current is switched off?

A coil like this is an **electro-magnet**.
An electromagnet usually has an iron **core**, which becomes magnetised and makes a stronger electromagnet.

You will investigate an electromagnet in the next lesson.

---

1 Copy and complete:
a) An electromagnet only works if a . . . . is flowing through the . . . .
b) The . . . . field round a straight . . . . is in the shape of circles.
c) The field round a coil (or . . . .) has the same shape as the . . . . round a bar . . . .

2 Make a survey of all the things in your house that use electromagnets.
(Hint: see the bottom of page 124.)

3 Drink cans are usually made from either steel or aluminium. In a metal re-cycling plant they need to be separated. Design a machine to do this.

4 In the diagrams below, **A**, **B**, **C**, **D** are compasses. In diagram (a), a current is flowing **down**, into the paper.
In diagram (b), there is no current flowing.
Copy the diagrams and draw in the direction of each compass needle.

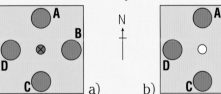

5 Plan an investigation to see if iron can be made into a magnet more easily than steel. How would you make it a fair test?

Things to do

8J4 Using electromagnets

Learn about:
- electromagnets
- uses of electromagnets

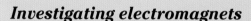

▶ Explain what is happening in this photo:

Is it an electromagnet or a permanent magnet?
In what ways are electromagnets and permanent magnets
a similar?
b different?

Investigating electromagnets

Plan an investigation to find out **what affects the strength of an electromagnet**.

- What factors can you vary?
  Choose **one** of these, and plan your investigation.

- How will you make it a fair and safe test?

- How will you measure the strength of your electromagnet?

Show your plan to your teacher, and then do it.
!Do not use electricity at more than 12 volts!

- If you have time, investigate the other variables.

Electromagnets have many uses.

An electric bell

This is used in doorbells, burglar alarms, and fire-bells.

Study the diagram:

c When the switch is closed, there is a complete circuit. What happens to the electromagnet?

The iron bar is on a springy metal strip which can bend.

d What happens to the iron bar?
e What happens to the hammer?

There is now a gap in the circuit, because the iron bar is not touching the contact.

f What happens now?

g Why does the bell ring for as long as the switch is closed?

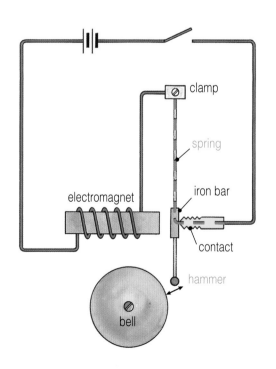

clamp

spring

iron bar

electromagnet

contact

hammer

bell

126

A relay

This is a switch operated by an electromagnet.
It is used when you want to use a small current
to switch on a larger current.

Study the diagram:

There are 2 circuits here.

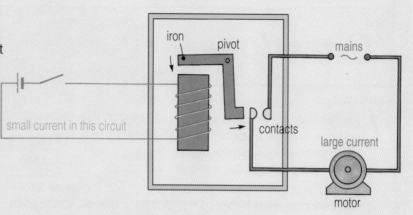

h When a current flows in the blue circuit,
what happens to the core of the coil?

i What happens to the iron bar?

j What happens to the contacts in the
red circuit?

k What happens to the motor?

The motor might be the starter-motor in a car, or
the motor in a washing-machine, or in an electric train.

A circuit-breaker

This is an automatic safety switch.
It cuts off the current if it gets too big.

Study the diagram:

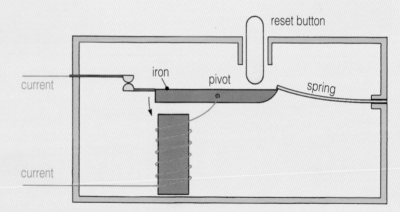

l What happens to the electromagnet when
the current is flowing?

m If the current is big, what happens to
the iron bar?

n What does this do to the current?

o How would you re-set this circuit-breaker?

p To make it switch off at a lower current,
how would you change the electromagnet?

Electromagnets are also used in loudspeakers, in motors, in cassette-recorders,
and in computers to store data on the computer discs.

---

1 Copy and complete:
The strength of an electromagnet can be
increased by:
a) increasing the number of turns on the . . . . ,
b) increasing the . . . . , or
c) using an . . . . core.

2 Design a relay that would use a small
current to turn **off** a big current. Draw a
labelled diagram of your design.

3 Door-chimes use an electromagnet to
make a 'bing-bong' sound. When the switch
is pushed, an iron rod hits one chime (a metal
tube). When the switch is released, the rod
springs back to hit the other chime tube.
Draw a labelled diagram to show how this
could work.

4 Design a machine that could separate full
and empty milk-bottles on a conveyor belt.

Things to do

Questions

1 Rachel tested 2 electromagnets, one with 20 turns and one with 60 turns on the coil. She counted how many nails each electromagnet could hold up at different currents. Here are her results:

Current (A)		0	0.5	1.0	1.5	2.0	2.5	3.0	3.5	4.0
Number of nails	20-turns	0	1	4	9	15	21	27	33	38
	60-turns	0	4	12	27	42	55	61	64	64

a) Plot a graph for each electromagnet (on the same axes).
b) Describe what happens when the current is increased in
  i) the 20-turn coil,   ii) the 60-turn coil.
c) Can you explain this?

2 Draw a diagram to show how a relay could be used to set off a large fireworks display safely.

Explain how it works.

3 In a hospital, the corridor doors need to be open most of the time, so trolleys can move easily. But in a fire, the doors should be closed to stop the fire spreading.

Draw a diagram to show how electromagnets can help.

4 The diagram shows a simple **loudspeaker**:

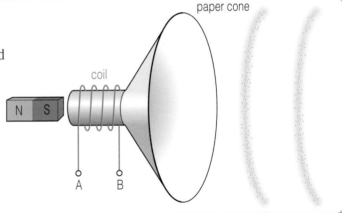

paper cone
coil
N S
A   B

If a current is passed through the coil from A to B, the coil becomes an electromagnet and is attracted to the bar magnet.

a) What happens if a current is reversed so it passes from B to A?
b) What happens if the current is reversed, to and fro, 50 times in each second? (This is called **alternating current**.)
c) What difference would you hear if the current was bigger?

5 Microwave oven
Microwaves are dangerous, and the oven must not work while the door is open.

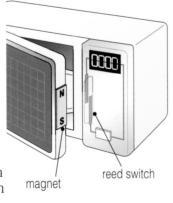

magnet   reed switch

a) Why are the 2 contact strips inside the **reed switch** made of iron?
b) Why does the oven not work when the door is open?
c) Why does the reed switch complete the circuit when the door is shut?

6 A door chime
Look at this circuit:

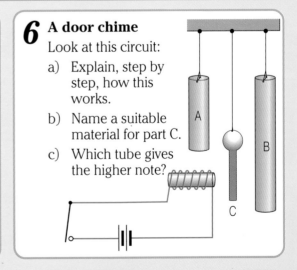

A
B
C

a) Explain, step by step, how this works.
b) Name a suitable material for part C.
c) Which tube gives the higher note?

Light

Light energy from the Sun is essential for life on Earth.

In this unit you can investigate light, including how we see things and why objects appear coloured.

See here!

Learn about:
● your eyes
● how you see things

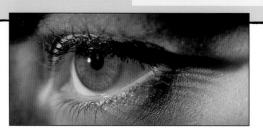

▶ Close your eyes and think what it would be like to be completely blind.
Write down 3 things that you could not do if you were blind.

▶ Hold a mirror in front of you, and look at your eye. You are looking at an **image** of your eye.
The coloured part of your eye is called the **iris**.
Make an accurate drawing of your iris.

The dark hole in the middle of your iris is called the **pupil**. This lets the light enter your eye.

▶ Keep looking in the mirror while you turn your head to point to a dark part of the room and then to a bright window.
Look carefully.
What happens to the size of your pupil? Why is this?

▶ Look at this diagram of your eye:
Study the different parts of it.

▶ Your teacher will give you a copy of this diagram. Fill in the missing words on the sheet.

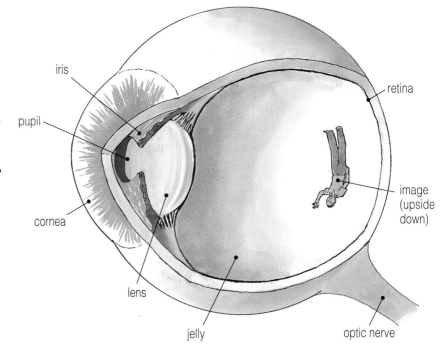

iris

pupil

cornea

lens

jelly

retina

image (upside down)

optic nerve

How do you see?

Emma has a hypothesis about how she can see things:

Tina has a different hypothesis:

● Which hypothesis do you think is correct?

● Write some sentences to explain your reasons.
Try to use these words:

light   reflected   eye   pupil

● Can you think of an investigation that will decide between these two hypotheses?

Emma says: "I think the light always has to travel to my eye and then to the book so that I can see it."

Tina says: "I think the light always has to travel to the book and then be reflected up into my eye."

Are two eyes better than one?

- Do you think you can judge distances better with one eye or two? Write down your prediction.

- Plan a short investigation to test this.
  You can use a quick test of skill such as: hold out a pencil in each hand at arm's length and try to make the pencil points touch each other.

- Do your investigation. What do you find?

Most animals have two eyes

Shadows

▶ Shine a torch or a **ray-box** at a white sheet of paper.
  Hold an object so that you see a **shadow** on the screen.

a Why is the screen bright?

b Why is the shadow dark?

c How can you make the shadow larger?

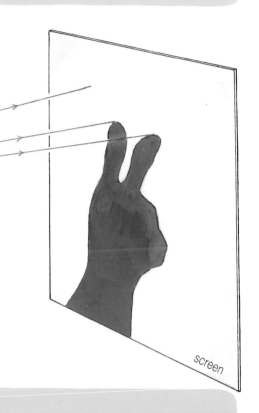

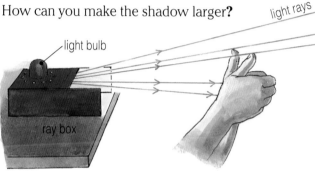
light bulb
light rays
ray box
screen

The light travels in straight lines. So we can draw straight lines called **rays** to show where the light is going.

Lighting up materials

Find out the meanings of all the words in the boxes:

| opaque   transparent   translucent | absorb   transmit   reflect | luminous   non-luminous |

Your teacher will give you a range of different materials.

- Use a ray-box, a light-sensor and data-logger, to classify them as opaque, transparent or translucent.
- Write about what you found, using enough sentences to explain all the words in the boxes.

---

1 Copy and complete:
When I look at this page, the . . . . rays are reflected off the white paper and then travel into my . . . . The rays make an . . . . on my retina. In a dim light my pupil grows . . . . so that more . . . . can get into my . . . .

2 In a thunderstorm you see the lightning and then hear the thunder. What does this tell you about the speed of light?

3 Draw a design for a clock that uses shadows from the Sun.

4 Get a piece of card about 10 cm × 10 cm. On one side draw a bird (using thick lines). On the other side draw a cage. Tape a pencil to the card so that you can spin it quickly between your hands. What do you see?
Write a sentence to explain what you think is happening. Then design a different card.

Things to do

Using mirrors

Learn about:
● the Law of Reflection
● images formed by mirrors

Light rays can be *reflected*.

You can see this page because light rays are being reflected off the white paper and into your eyes.

Where the light shines on this black ink, it is not reflected – it is **absorbed**.

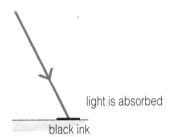

light is reflected to your eye

white paper

light is absorbed

black ink

A mirror is a good reflector of light.
When you look in a mirror, you see an **image** of yourself.

▶ Think about all the ways that mirrors are used – in homes, shops and cars. Make a list of all the uses you can think of.

How can you use a mirror on a sunny day to send a message to a distant friend?

If your bedroom is dark, how can you use a mirror to make it look brighter?

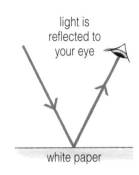

A flat mirror is called a **plane** mirror.
Here is an experiment to see what happens when light is reflected off a plane mirror.

The Law of Reflection

Your teacher will give you a Help Sheet with some lines and angles marked on it.

1 Set your plane mirror with its **back** along the line marked 'mirror' (the light is reflected from the silver back of the mirror).

2 Use a ray-box to send a narrow beam of light (a 'ray') along the line marked 20° on the **angle of incidence** scale.

3 Measure the **angle of reflection**. What do you find?

4 Repeat this using different angles of incidence, and record your results.

5 Draw a line-graph to display your results, with a line of best fit through your points.

6 Write down your conclusion.

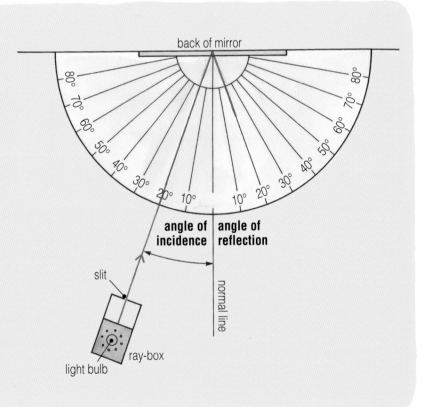

back of mirror

angle of incidence | angle of reflection

slit

normal line

ray-box

light bulb

Mirror images

1 Fix a sheet of glass so that it is upright on the table.

2 Put a Bunsen burner with a bright yellow flame in front of it.

3 Look into the 'mirror' to see the image of the flame. Where does it appear to be?

4 Move another **un**lit Bunsen burner until the image of the flame sits exactly on it.

5 Measure the distances of the two Bunsen burners from the glass mirror. What do you find?

6 Try this using different distances.

7 Write down your conclusion.

• What do you see if you put your finger on top of the **un**lit Bunsen?

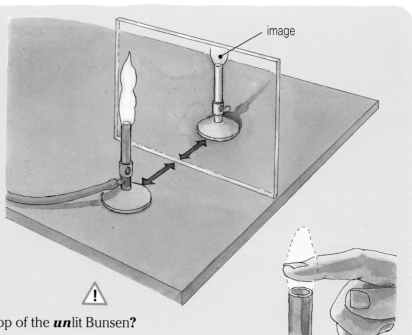

image

Curved mirrors

Curved mirrors can be **convex** (like the back of a spoon) or **concave** (like the front of a spoon).

A **concave** mirror is used in a torch and in a car headlight:

a concave mirror in a torch

light rays

light bulb

A **concave** mirror is also used in a solar cooker:

This is a cheap source of energy in some countries.

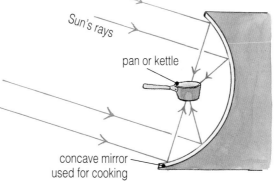

Sun's rays

pan or kettle

concave mirror used for cooking

---

1 Look at this notice:

The Law of Reflection
The angle of incidence is **equal** to the angle of reflection

a) How can you read it easily?
b) Copy it out correctly.
c) Copy and complete:
    The distance from an object to a plane mirror is . . . . to the distance from the . . . . to the mirror.

2 Write your name so that it reads correctly when viewed in a mirror.

3 Where and why might you see:
    AMBULANCE

4 Imagine you wake up tomorrow in a world where light is never reflected. Write a story about it.

5 Read the next two pages (8K3) and decide what you need to bring to the next lesson.

Things to do

Using light rays

Choose either   • the **pin-hole camera** (below)
          or   • the **periscope** (opposite page)
and then build it.

Making a pin-hole camera

▶ Look at the diagrams and then decide
how to make your camera.

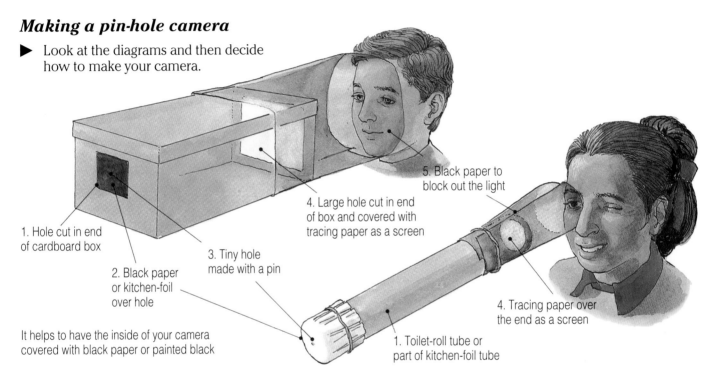

5. Black paper to
block out the light

4. Large hole cut in end
of box and covered with
tracing paper as a screen

1. Hole cut in end
of cardboard box

3. Tiny hole
made with a pin

2. Black paper
or kitchen-foil
over hole

4. Tracing paper over
the end as a screen

It helps to have the inside of your camera
covered with black paper or painted black

1. Toilet-roll tube or
part of kitchen-foil tube

▶ Use your camera to look at a light bulb or out of the window.

a What do you see?

b Which way up is the *image*? (We say it is *inverted*.)
Can you explain why?
(Hint: think of the light rays coming in through the hole.)

c How could this camera be used by an artist to make a painting?

d If you wanted to take a photo, where would you put the film?

e What happens if you make the pin-hole wider? Try it.
Is the image brighter or darker? Why?
Is the image sharper or more blurred? Why?

f Explain how your camera works, using the words:

> **light rays   pin-hole   straight lines   image**

g How could you make the image twice as big?

• If you have time, make the pin-hole much bigger and then
put a lens over the hole. Move the lens until you get a sharp
and bright image. You have made a lens camera.

A photo taken with a pin-hole camera

Making a periscope

▶ Look at these diagrams and then make your own periscope.

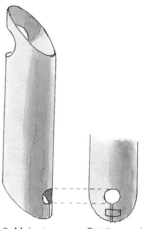
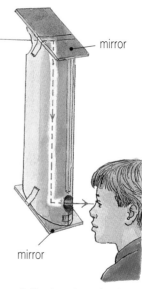

mirror

mirror

1. Roll of cardboard (or the tube from the centre of a kitchen-foil roll, or the box it is sold in, or two/three toilet-role tubes).

2. Cut off the ends, at exactly 45°, as shown.

3. Make two viewing holes as shown.

Cut the card, then sellotape afterwards.

The bottom hole can be small but make the top one as big as you can.

3. Put the mirrors in place and hold them with sellotape.

Adjust the mirrors – and the viewing holes – until you can see through your periscope properly.

▶ Use your periscope to look over an object (the table or the window-sill).

h What do you see?
You are looking at an *image* in the mirrors.

i Describe the image that you see. For example:
 - Is it the same way up as the real thing?
 - Is it the same size as the real thing?
 - Is it the same colour as the real thing?

j How can you use your periscope to see round a corner?

k How many different uses for a periscope can you think of? Make a list.

l How could you change the design so that you could see backwards over your head? Sketch a design for this.

m Explain how your periscope works, using the words:

light rays	*straight lines*	*top mirror*
plane	*bottom mirror*	*eye*	*image*

---

1 Copy and complete:
In a pin-hole . . . . , the light . . . . enter through the . . . . and travel in . . . . lines to the screen where they form an . . . .
The image is . . . . (upside-down).
If the pin-hole is made . . . . , the image is brighter but more blurred.

2 Explain how you can convert a pin-hole camera into a lens camera. What is the advantage of doing this?

3 Copy and complete:
In a periscope, the light . . . . enter the top hole and are . . . . by the top mirror, so that they travel . . . . the tube to the bottom . . . . , where they are . . . . into your . . . . , so that you can see an . . . .

4 A driver is using her car to pull a caravan, but it blocks her view through the driving mirror.
Design a periscope to solve the problem, and draw a labelled sketch.

Things to do

Mirror ɿoɿɿiM

Learn about:
● reflection of light rays
● uses of mirrors
● multiple images

▶ Kate is looking into a **plane** mirror.
A ray of light from the lamp is *reflected* from the mirror:

a Which is the incident ray? Which is the reflected ray?

b If the angle of incidence is 20°, how big is the angle of reflection?

c Explain why Kate sees the lamp.

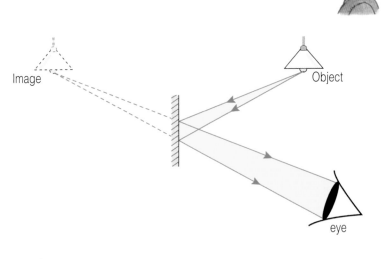

20°

▶ Kate sees an **image** of the lamp. It is called a *virtual* image – you cannot touch a virtual image.

d Write down the word IMAGE as it would look when seen in a mirror.

▶ Here is another diagram of Kate looking at the lamp:

It shows 2 rays from the lamp going into Kate's eye.
When Kate looks at the mirror, she sees the image **behind** the mirror.
The image is *where the rays appear to come from*.

e If the lamp is 2 metres from the mirror, where exactly does Kate see the image?

Image

Object

eye

▶ This diagram shows a beam of light being *scattered* from a piece of paper:

f Why can't you see an image in a sheet of paper?

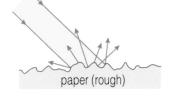
paper (rough)

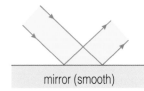
mirror (smooth)

MIRROR ЯOЯЯIM – 1

Your teacher may give you a Help Sheet with these diagrams:

Tina likes to go to pop concerts, but often she can't see over the crowd.
How can she use mirrors to see the band?

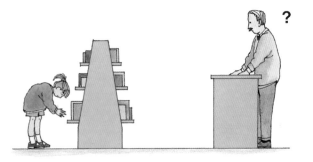
Mr Brown wants to see all the shelves in his shop, in case of shop-lifters.
How can he use a mirror (or 2 mirrors) to see his shelves?

MIRROR ЯОЯЯIM – 2

Your teacher may give you a Help Sheet with these diagrams:

Round the bend

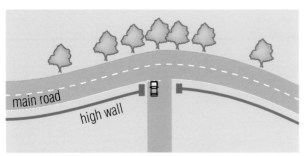

This plan shows a dangerous road. A car-driver in the side-road cannot see traffic on the main road. How could you use 2 mirrors to help the driver?

Mirror Island

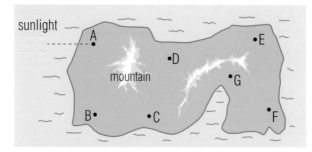

This map of an island shows sunlight arriving at A. How could you use mirrors to reflect the sunlight to G? How could a person at A send messages to a friend at G?

Multiple images

Place 2 plane mirrors at an angle, with an object between them:

Look past the object. How many images can you see?

Investigate how the number of images depends on the angle between the mirrors.

- How can you arrange to place the mirrors accurately, at a specific angle to each other?

- Try angles of 90° and 120° at first. Which other angles are 'good' angles?

- A full circle is 360°. Can you deduce a formula for the connection between the **angle** and the **number of images?**

- What is the name of the toy that uses this idea? Can you make one? What is the best number of mirrors to use?

sharp edges

Infra-red

Most TVs have a remote-control handset, that gives out ***infra-red*** rays. Infra-red rays travel at the same speed as light, but we cannot see them.

Plan an investigation to see if infra-red rays are reflected in the same way as visible light rays.

If you have time, try it.

---

1 Copy and complete:
a) When light is reflected, the angle of . . . . is . . . . to the angle of . . . .
b) The distance from an object to a plane mirror is . . . . to the distance from the . . . . to the mirror.
c) The image in a plane mirror is called a . . . . image.

2 Think about all the ways that mirrors are used – in homes, shops, and cars. Make a list of all the uses that you can think of, in 2 columns: (1) plane mirrors, (2) curved mirrors.

3 In your house, where would you put some mirrors so that you could see who is at the front door while you are lying in bed.

Things to do

Bending light

Learn about:
● refraction
● different types of lens

▶ Look at this picture:

A ray of light from the lamp is **reflected** from the mirror.

a Which is the incident ray?

b If the angle of incidence is 20°, how big is the angle of reflection?

c Why does the boy see the lamp?

d Where does the image of the lamp appear to be?

Reflection from a mirror is one way of changing the direction of a ray. Here is another way, using **refraction**.

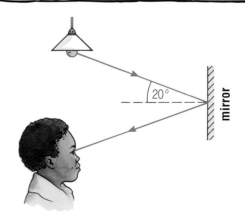

Investigating refraction

Your teacher will give you a Help Sheet and a semi-circular block of glass (or perspex).

1 Place your block of glass in position on the Help Sheet, as shown here:

2 Use a ray-box to send a thin beam of light at an angle of incidence of 30°, as shown.

3 Look carefully at the ray where it comes out of the block.
Can you see the ray does not go straight on? It changes direction at the surface of the block. We say the ray is **refracted**. This is **refraction**.

4 Mark on the paper the path of the ray coming out of the block. Label it ray ①.

5 Now increase the angle of incidence to 40°. What happens?
Mark the new path of the refracted ray, and label it ②.

You can see that when light comes out of a glass block, it is bent (refracted) **away from** the normal line.

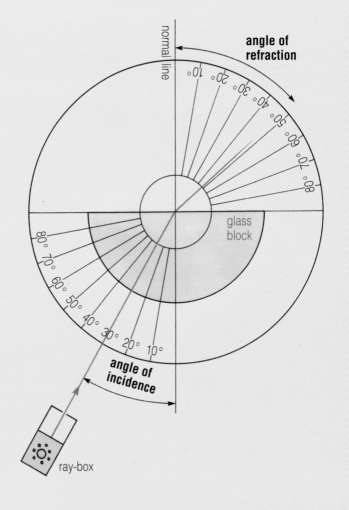

6 Now increase the angle of incidence to 50°. What happens now?
Mark the new path of the ray, and label it ③.

When the angle of incidence is large, you can see that the light is reflected **inside** the glass. We call this total internal reflection.

Refraction in a rectangular block

▶ The diagram shows a ray of light going into a glass block:

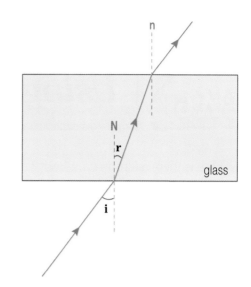

e What happens to the ray when it enters the glass?
On the diagram you can see a blue dotted line labelled **N**.
This is called the **Normal** line.

f Is the light ray bent away from or towards the normal?

g Which is bigger – the angle of incidence (**i**) or the angle of refraction (**r**)?

Light travels very fast in air – at 300 000 km per second!
In glass it travels more slowly. As the light is slowed down,
it is refracted towards the normal.

h What happens to the light ray as it leaves the glass?

An 'imagic' trick

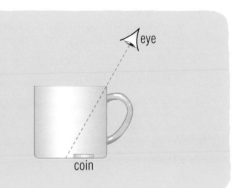

Put a coin at the bottom of an empty cup. Move your head until
the coin is **just** out of sight.
Without moving your head, pour water into the cup (or ask a friend
to pour). Do it slowly so that the water does not move the coin.

• What do you see?

• Can you explain this, using refraction?
Can you draw a ray diagram to show why it happens?

Using refraction

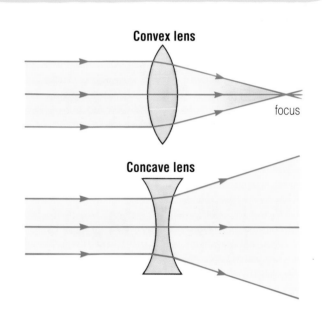

A **lens** is a shaped piece of glass. There are two kinds:

A **convex** lens is fat in the middle.
A **concave** lens is thin in the middle.

When light goes through a lens it is refracted.

A convex lens brings the rays of light closer together.
We say they are **converging**.

A concave lens makes the rays spread out. They are
diverging.

The rays always bend towards the thickest part of a lens.

i Where is there a lens in your body?

j Is it convex or concave?

---

1 Copy and complete:
a) When light goes in or out of glass, it . . . .
direction. The rays are . . . . (bent).
b) When light comes out of glass it is . . . .
away from the normal line. When it goes
into glass it is . . . . towards the normal.
c) In a convex lens the . . . . come closer
together. The rays are . . . .

2 Look at the diagram of a **convex** lens on
this page.
Draw similar diagrams to show what you
think would happen to the rays if the lens
was: a) fatter, b) thinner.

3 Make a list of all the things you can
think of that use a lens.

Things to do

Using lenses

▶ Where is there a lens in your body?
Is it convex or concave?
Does it converge or diverge rays of light?

▶ Sketch a simple diagram of an eye, like the one shown.
Then add these labels in the right places:

> **lens   retina   iris   pupil   cornea**

a You are using your eyes to see this page.
Explain, step by step, how the light travels from the window until it is focussed on your retina.
You can start like this:

Light from the window shines on the book, and then...

b What happens to the pupil in your eye if you look at a bright light?

Focussing your eyes

The lens in your eye can change shape. When you look at near objects it gets fatter. For far objects it gets thinner.
The muscles in your eye make the lens go fatter or thinner, until the image is sharp:

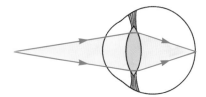

Looking at a near object, your lens is *fat*.

If you can't see a near object clearly, you are long-sighted.
(You may need spectacles with con<u>vex</u> lenses.)

Looking at a far object, your lens is *thin*.

If you can't see a distant object clearly, you are short-sighted.
(You may need spectacles with con<u>cave</u> lenses.)

Eye tests

Plan, and carry out, an investigation to find out the distances at which you can read letters of different sizes.

● How will you make it a fair test?

● Is it the same for the left eye, the right eye, and both eyes?

● Plot a graph of the **distance from your eye** against the **size of the letter**. What do you find?

● Is your graph the same as other people's?

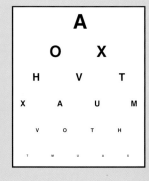

A	4 mm
O X	3 mm
H V T	2 mm
X A U M	1.5 mm
V O T H	1 mm
T M U A X	0.8 mm

size of letters

The camera

In a camera, a lens is used to make an image on the film:

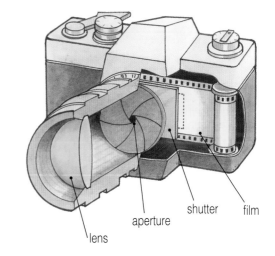

Use a convex lens to focus the light rays from a lamp, like in a camera:

- What do you notice about the image?
- Move the lamp to different distances from the screen. Each time, focus the image. Measure the distances shown on the diagram, and record them in a table.
- What pattern do you find?

- In a camera, how do you focus on
  - near objects?
  - far objects?

- Does your eye focus on objects in this way?

▶ Look at this diagram of a camera:

c Which part of the camera is like your retina?

The **aperture** can be changed to let in more or less light.

 open   closed

d Which part of your eye is like this?

e When should the camera use a small aperture?

f In what other way can a camera change the amount of light going to the film?

g The camera and your eye both use a lens. In what ways are the lenses i) similar? and ii) different?

h Explain carefully how your eye and a camera use different ways to focus the image.

---

1 Copy and complete:
a) My eye lens is a . . . . lens. It . . . . the rays of light.
b) To focus on near and far objects, my eye . . . . changes shape. To focus on . . . . objects, it is fatter.
c) A long-sighted person cannot focus on . . . . objects. A short-sighted person cannot focus on . . . . objects.
d) The . . . . in a camera and in my eye are inverted (upside-down).

2 Copy and complete:
a) A camera uses a . . . . lens.
b) To focus a camera on near objects, the lens is moved . . . . from the film.
c) The . . . . in a camera is like the iris in my eye.

3 Draw up a table or a poster which shows all the ways in which a camera and your eye are i) similar, and ii) different.

 Things to do

A world of colour

Learn about:
● making a spectrum
● light waves
● seeing coloured objects

a Why do you think road signs are often coloured red?

b Imagine a world without colour. Describe what it would be like to live in it.

c Write down what you think are the colours in a rainbow.

The colours in a rainbow form a **spectrum**.

Making a spectrum

Shine some white light from a ray-box through a prism, and on to a screen:

Turn the prism until you see a spectrum on the screen.

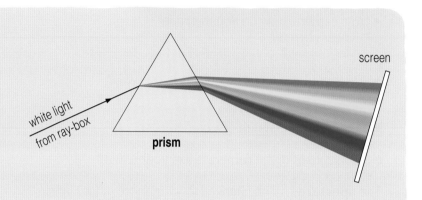

d How many different colours can you count?

e Which colour has been bent the least?

f Which colour has been refracted the most?

This experiment shows that white light is really a mixture of several colours. The colours are split up by the prism.

We say that the white light has been **dispersed** by the prism to form a visible spectrum. This is **dispersion**.

The colours, in order, are: **R**ed, **O**range, **Y**ellow, **G**reen, **B**lue, **I**ndigo, **V**iolet.
You can remember them as a boy's name: **ROY G. BIV**.

Can you make up a sentence to remember the colours?
(e.g. Really Old Yachts … or Rolling On Your …)

Combining the colours

Can you find a way to re-mix the colours of your spectrum?

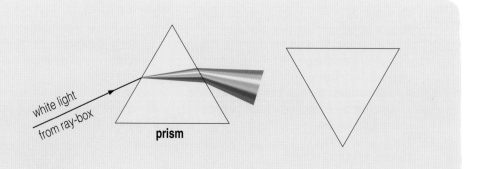

g What colour do you get?

142

Light waves

If you throw a stone in a pond, you can see the ripples or **waves** spreading out from it.
Light spreads out in the same way, in waves.

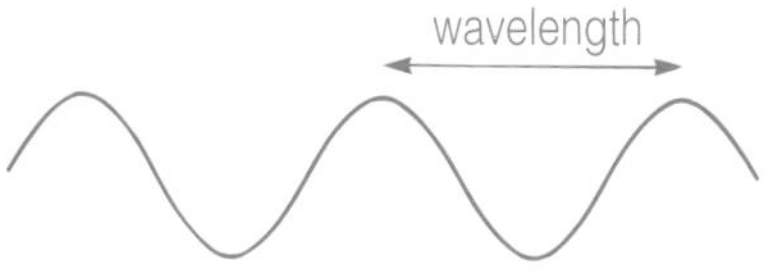

Each wave has its own **wavelength**.
Different colours have different wavelengths.

Red light has the longest wavelength. It is about $\frac{1}{1000}$ mm.

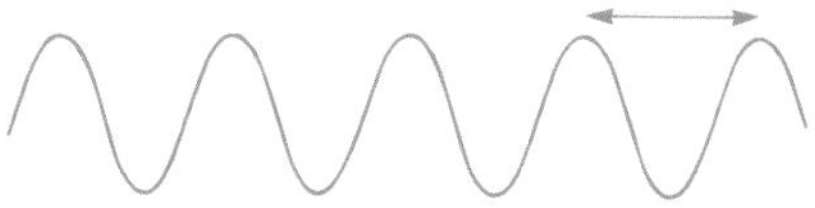

Violet light has the shortest wavelength. There are about 2000 wavelengths of violet light in 1 mm.

The light is refracted as it enters the prism.

The light is refracted because it travels slower in the glass than in the air.
Different colours travel at different speeds, and so are refracted by different amounts.

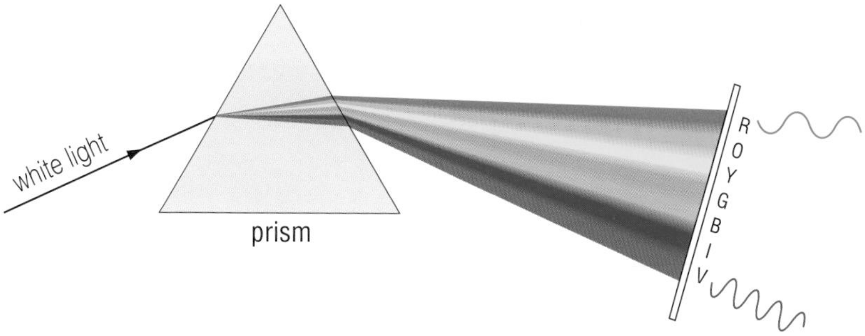

h Which colour is refracted most**?**

i Which colour do you think travels slowest in glass**?**

Seeing coloured objects

When light shines on a coloured object, some of the light is taken in or ***absorbed***.
The rest of the light is reflected.
You see the colour of this reflected light. For example:

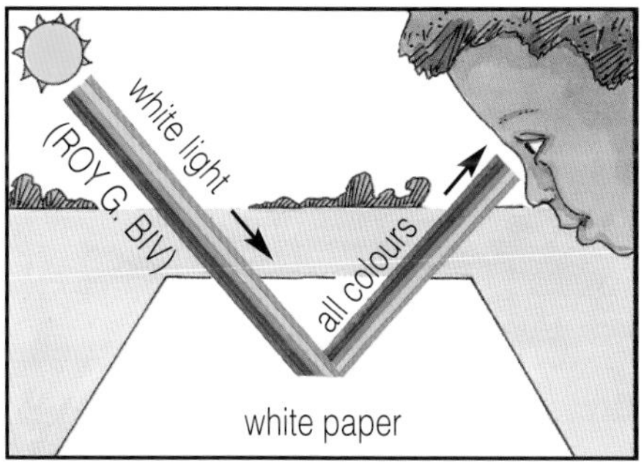

White things reflect all of the colours of light. That is, all of ROY G. BIV.

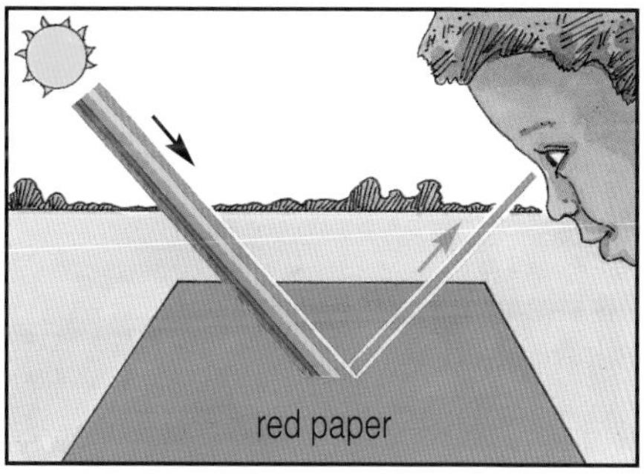

Red things reflect red light and absorb the other colours. We see the red light.

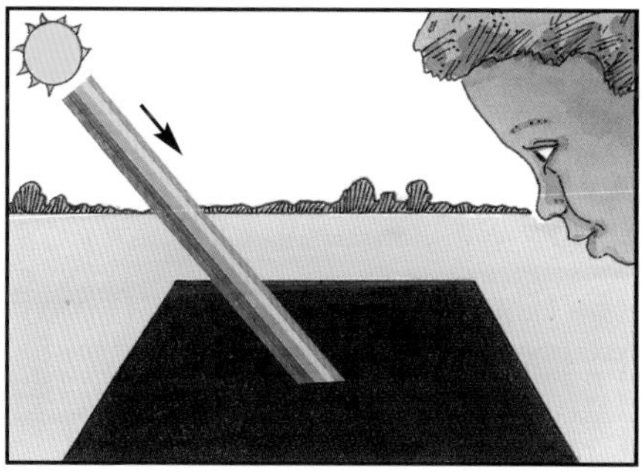

Black things do not reflect any light. All the light is absorbed.

j What colour light is reflected by a blue T-shirt**?**

k Explain what is happening when you look at this red ink.

Things to do

1 Copy and complete:
a) . . . . light is a mixture of 7 colours.
b) The 7 colours of the spectrum are:
c) . . . . light has the longest wavelength.
d) . . . . light is refracted the most.
e) A red T-shirt reflects . . . . light and . . . . all the other colours.
f) Black ink does not . . . . any light.

2 Explain the following:
a) Why does the paper appear white?
b) Why does this blue ▬ look blue?
c) Why does this ink look black?

3 What is camouflage? Draw some camouflage suitable for a bird-watcher
a) in the desert, b) in the jungle.

More about colour

▶ Which is your favourite colour?
What does it remind you of?

a How many colours are there in a rainbow?

b How can you remember their names?

c How can you produce a spectrum?

The visible spectrum

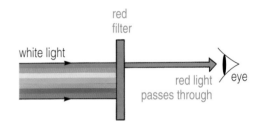

Can you see a red number here?
If you can't, you may be colour-blind.

Filters

What do you see when you look through a piece of red plastic or red glass?
A red **filter** will only let through red light:
It **absorbs** all the other colours:

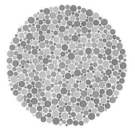

d Which colour passes through a green filter?

e Which colours are absorbed by a green filter?

f Use this diagram to explain what you see when you look through a blue filter and a red filter together:

Use these words in your answer:

> **transmit**   **absorb**

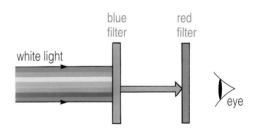

g Where have you seen filters used?

Mixing coloured lights

Red, **Green** and **Blue** are called the **primary** colours of light.
(These are not the same as the primary colours used in painting.)

When 2 primary colours of light overlap, you get a **secondary** colour:
These are called   yellow ,   cyan   and   magenta .

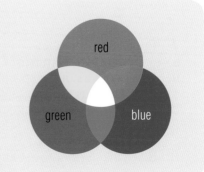

h Which 2 colours give yellow?

i What colour do you get when all 3 primary colours of light overlap?

Looking at objects in different colours of light

Plan an investigation that will give you the data for this table:

- To make the coloured light you can use a coloured filter on a ray-box (or torch). A red filter lets only red light through it.

- Show your plan to your teacher, and then do it.

- What pattern do you find?

Colour of objects in coloured lights

Colour of object in daylight	Colour of light shining on it			
	white	red	green	blue
white	white			
red				
green				
blue				

Designing for the stage

Imagine you are the stage-designer for a pop group. You have to design the band's clothes as well as the stage colours. The manager tells you that the stage lights will flash red or green or blue.
The picture shows someone's first attempt:

- Look at this picture in red light, in green light, and in blue light. (Or look through filters.)

- Then re-design the set and the clothes so that the band can be seen better.

- Explain why it is often difficult to decide the colour of cars or clothing under street lights.

Safety first

Plan an investigation to see which is the safest colour for you to wear when riding a bicycle, or walking on a road.

- How will you make it a fair and safe test?

- How will you find out which is the safest colour for both day-time and night-time?

- Show your plan to your teacher and, if you have time, do it.

- How else can you improve your safety on the road?

1 Copy and complete:
a) White light is . . . . by a prism into a spectrum. The 7 colours are:
b) A red filter lets . . . . light through, and . . . . all the other colours.
c) A blue T-shirt reflects . . . . light, and . . . . all the other colours.

2 What colour would a red book look:
a) in white light?   d) in blue light?
b) in red light?   e) through a red filter?
c) in green light?   f) through a blue filter?

3 Which jobs may be dangerous or difficult if you are colour-blind?

Things to do

Physics at Work

Road safety

In the photo, the coat and the road-sign contain thousands of tiny shiny beads that act like mirrors.

a Explain why the man's coat looks brighter than his face.

b What does this say?

c Why is it written like this?

d How should the word STOP be written on a sign on the front of a police car?

Use these 3 ray diagrams to explain,
in your own words,
and with examples,
the meanings of:

e transparent,

f translucent,

g opaque.

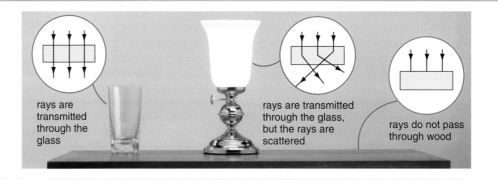

rays are transmitted through the glass

rays are transmitted through the glass, but the rays are scattered

rays do not pass through wood

Autocue

Politicians and newsreaders often use an autocue.
It lets them talk directly at the audience, so they don't have to look down at their notes, or memorise them.

h Use the diagram to explain why the speaker can see the words.

i Explain why the audience can't see the words.

j Look at the way the words are written on the projector. Why is this?

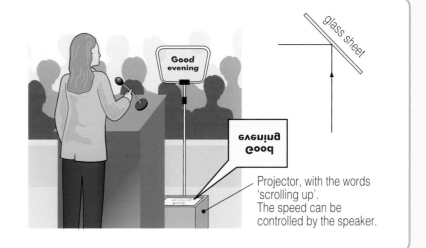

glass sheet

Good evening

Good evening

Projector, with the words 'scrolling up'. The speed can be controlled by the speaker.

Red + Green + Blue

The picture shows 3 spotlights shining on a screen:

k What colour do you get where the red + green overlap?

A **TV screen** has tiny spots of red, green and blue on it.

l What is happening when you see yellow on the TV screen?

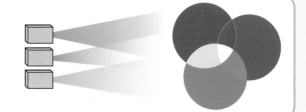

Ideas about light

2500 years ago **Pythagoras** suggested that he saw an object because it gave out a stream of particles which travelled to his eyes, like a stream of tiny bullets. The particles were called '***corpuscles***'.

2200 years later, in 1666, **Isaac Newton** investigated white light passing through a prism. He showed that white light is made up of 7 colours (ROY G BIV).

In 1675, **Olaus Romer** measured the speed of light. It travels at 300 000 km per second! This is about a million times faster than the speed of sound.

About the same time, **Christiaan Huygens** suggested that light travels as a ***wave***, not as corpuscles.

OR

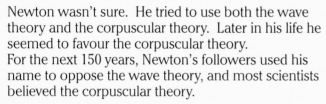

waves spreading out like ripples on a pond?

corpuscles shooting out like bullets?

Sir Isaac Newton

Newton wasn't sure. He tried to use both the wave theory and the corpuscular theory. Later in his life he seemed to favour the corpuscular theory.
For the next 150 years, Newton's followers used his name to oppose the wave theory, and most scientists believed the corpuscular theory.

Then in 1801, **Thomas Young** found convincing evidence for the wave theory. So for the next 100 years this theory was believed.

Then in 1905, **Albert Einstein** found good reasons to suggest that light travels as bundles of energy called ***photons***, rather like corpuscles.

Today, physicists believe that light can behave as both waves ***and*** particles. Sometimes it behaves more like a wave, sometimes more like a particle!

Albert Einstein

1 A light-year is a distance. How far is it?

2 Why do you see lightning before you hear the thunder?

3 Plan an investigation to see if snooker balls (or tennis balls) are reflected off a wall following the same rule as for light rays. Draw a diagram and explain exactly what you would do.

4 Explain how Newton's reputation was used in a non-scientific way to support the corpuscular theory.

5 The Sun is 150 000 000 km from Earth. How long does it take the sunlight to reach your face?

6 Choose one of the scientists and find out more details of his life.

Things to do

Questions

1 Look at your pen. Explain how you are able to see it.

2 Suppose you can choose from a variety of mirrors (plane and curved) and lenses (converging and diverging).
Work out a design for each of these:
a) a torch
b) a periscope to see at a football match
c) a periscope to see behind you in a car
d) a solar cooker
e) a spy camera to take photos round corners
f) an over-head projector for a teacher.

Draw a labelled diagram of each design. On each diagram, draw coloured lines to show what happens to the rays of light.

3 Natalya is only 3 years old. She can speak but she can't read yet, so it is hard to use the usual eye-test with her:
Design a test that could be used instead.

4 Some pupils were hypothesising about the effects of colour.

Anna said, "I think more people choose to eat green jelly than any other colour."
Jamie said, "I think that flies are more likely to land on yellow surfaces than white surfaces."

Choose one of these hypotheses, and plan an investigation to test it. Take care to make it a fair test.

5 The diagram shows the path of 2 rays of light from a fish to a fisherman:

a) Use the diagram to explain why the fisherman should not aim his spear directly where he sees the fish.
b) Why does water always appear shallower than it really is?

6 Can you find 6 consecutive letters in the alphabet that look the same when a mirror is placed to reflect half the letter?
Which other letters are symmetrical?

7 A bike reflector is made of triangular pieces of red plastic:
The back of the plastic acts as a mirror.
a) Draw a diagram to show exactly how it reflects the light from the headlights of a car.
b) Why is the reflected light coloured red?

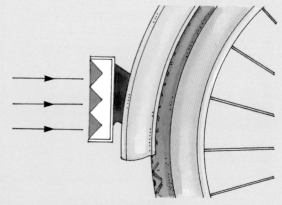

8 Plan an investigation to see which colours are best for an easy-to-read disco poster.

Sound and hearing

8L

We use our ears all the time.

We need them to make sense of the world around us.

Sound waves are useful to us in many other ways as well, as you will discover …

Sound moves

Learn about:
● changing pitch, loudness
● the speed of sound
● echoes

▶ Sit quietly and just **listen**. Make a list of all the sounds you can hear in one minute.

▶ Write down as many 'sound' words as you can. For example, boom, bang, crash, squeak, . . .

▶ Touch the front of your throat while you make an 'aaah' sound. Can you feel it **vibrating**?

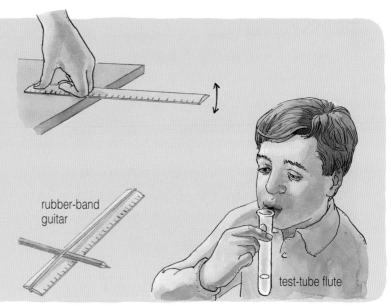

Hold a ruler firmly with part of it over the edge of the table. Then twang it.

a What is the end of the ruler doing?

b When does it stop making a sound?

c How can you make the sound quieter? How can you make it louder?

d How can you make it sound a higher note? And then a lower note? Can you play a tune – for example, 'Jingle Bells'?

Now repeat steps **a** to **d** with the other two 'musical instruments' shown here:

What do you find? Can you see any patterns? Write down your conclusions.

rubber-band guitar

test-tube flute

To make the ruler vibrate, you had to give it some **energy**.

e Where did this energy come from?

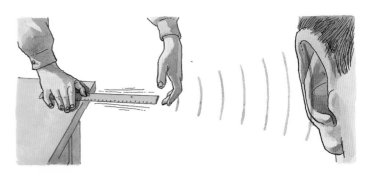

The vibrating ruler sends out sound waves through the air. Some of this sound energy travels to your ear, and so you can hear the sound.

Sound can travel a long way. In a thunderstorm, you can see a flash of lightning and then later you can hear the sound of it (the thunder).

f Which travels faster, light or sound?

Sound travels about 330 metres in one second. (Light travels almost a million times faster, at 300 000 000 metres per second.)

g How far would sound travel in 2 seconds?

h If you hear thunder 10 seconds after the lightning flash, how far away is the storm?

Echoes

If you clap your hands in front of a big building, you may hear an **echo**.

This happens because the sound wave is *reflected* back to you. The building is like a mirror.

Suppose you heard the echo after 2 seconds.

sound wave travels to the wall

clap

and back again

i How long did it take for the sound to get *to the wall?*

j How far away is the wall, if the speed of sound is 330 metres per second?

Echo-sounding

Sailors can use echoes to find the depth of the sea, using an *echo-sounder* or *sonar*.

Suppose this ship sent out a sound wave, and it got back an echo after 1 second.

k How long did it take the sound to get to the bottom of the sea?

Sound travels faster in water. It travels at 1500 metres per second.

l How far will the sound travel in $\frac{1}{2}$ second?

m How deep is the sea under the ship?

n If the shoal of fish in the diagram swims under the boat, how will the captain know?

The sound used by the sonar is too high for us to hear. It is called **ultrasonic** sound or **ultrasound**.

Dolphins use ultrasound to find their food. They make high-pitched squeaks and listen to the echoes.

Bats also use ultrasonic sounds, so that they can find food and 'see' in the dark

▶ Plan a safe investigation to find the speed of sound.
  • What equipment would you need?
  • What measurements would you take?
  • How would you calculate the speed?

---

1 Copy and complete:
a) All . . . . are caused by vibrations.
b) Echoes are due to the . . . . of sound.
c) The speed of sound in air is 330 . . . . per second.

2 Think about the noise in your school dining-hall. Write a list of suggestions for making it quieter.

3 Watching a cricket match from a distance, it seems that the bat hits the ball before you hear it. Explain this.

4 Karen hears an echo from a cliff after 4 seconds. How far is she from the cliff?

5 Write a poem using as many 'sound' words as possible.

Things to do

Hear, hear!

Learn about:
● how we hear sounds
● investigating our hearing

▶ Look at this diagram of your ear:
Study the different parts of it.

vibrating ruler ↑

sound waves travelling to your ear

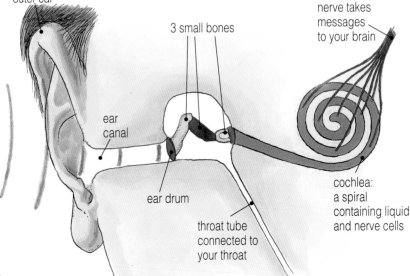

outer ear

3 small bones

nerve takes messages to your brain

ear canal

ear drum

throat tube connected to your throat

cochlea: a spiral containing liquid and nerve cells

The vibrating ruler is making some sound waves.
These waves travel through the air and make your
ear-drum vibrate.
This makes the 3 small bones vibrate, and they make the liquid in
your **cochlea** vibrate. This affects the nerve cells in the cochlea and
a message is sent to your brain . . . and so you hear the sound.

▶ Your teacher will give you a copy of this diagram.
Fill in the missing words on it.

Looking after your ears

Your ear is very delicate and can easily be damaged.
This damage can cause deafness.

● Your ear canal can become blocked with wax.
If so, the doctor can wash it out.

● Your ear-drum can be torn by a very loud bang, or
damaged by an infection. It may mend itself, or
doctors can graft a new one.

● Your ear bones may stick together and stop
vibrating properly. An operation can fix this.

● Your 'middle ear' (the small bones and the throat
tube) may be infected. Antibiotics can cure this.

● Your cochlea can be damaged by loud noises – for
example, at pop concerts, near noisy machines, or
wearing 'walkman' headphones. There is **no** cure!

As people get older, their ears work less well. Partial
deafness can be helped by wearing a **hearing-aid**.
This **amplifies** the sound to make it louder.

Wearing 'ear defenders' at work

How does the size of your outer ear affect your hearing?

Investigate whether the size of the outer ear affects how well you can hear faint sounds.

You can make yourself larger 'ears' from card.

- What faint sounds will you try to hear?
- How can you carry out a *fair and safe test* of different ear sizes?
- How will you record your results?
- How can you make your results more reliable?

Do your investigation. What do you find?

Does the *size* of the ear matter, or the *shape* of the ear, or both?

Sketch the shapes you have used.

Look at this bush-baby.
What conclusions can you make?

Are two ears better than one?

Investigate whether someone can tell the direction of a sound better with two ears or one.

- How will you make sure the person being tested uses only their ears?
- What will you use to make a sound?
- How will you make it a fair and safe test?
- How will you record your results? Can you record them on a diagram or a map?

Show your plan to your teacher and then do it.

What do you find? Which is better: one ear or two?

Are people more accurate in some directions than others?

Write a report explaining what you did and what you found.

Evaluate your evidence. Is it reliable?

1 Someone plucks a guitar string and you hear it. Explain, step by step, what happens between the guitar string and your brain.

2 Design a poster to encourage teenagers to look after their ears better.

3 People sometimes cup their hands behind their ears when they are trying to hear. Explain why this helps, using all these words if you can:

> vibrations   sound waves
>   reflection   like a mirror
> concave   sound energy   ear

Things to do

Sound waves

Learn about:
- how sound travels
- frequency of vibrations
- different hearing ranges

▶ Look at the picture:

a Jan's throat is vibrating. Explain, step by step, how Mei hears the sound waves.

▶ Rabbits warn each other of danger by thumping the ground.

b Do you think sound can travel through a solid?

c How do whales 'talk' to each other?

d Can sound travel through water?

Sound travels

Use a 'slinky' spring to show how sound moves.

Push one end of the spring to compress it:

vibration

Then pull it back:

Then push it again, to compress it:

wave
This kind of wave is called a longitudinal wave.

You are **vibrating** the end of the spring.
A **wave** of energy travels down the spring.

You can see that some parts of the spring are pushed together, and other parts are pulled apart.

When you speak, sound energy travels from your mouth in the same way.
Instead of a spring, there are air particles called molecules.
When you speak, the molecules are pushed together and pulled apart, so that they vibrate like the spring.
The sound energy travels away from your mouth, like the wave on the spring. Sound travels 330 metres in every second.

Can sound travel through a vacuum?

e What is a vacuum?

Your teacher will show you what happens to sound when the air is removed from a bell-jar:

f First, **predict** what you expect to happen.

g What happens as the air is pumped out?
h What happens as the air is let back in?
i Explain this experiment.

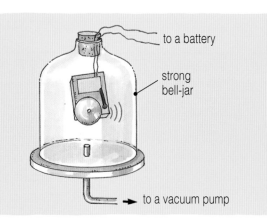

to a battery

strong
bell-jar

to a vacuum pump

What notes can you hear?

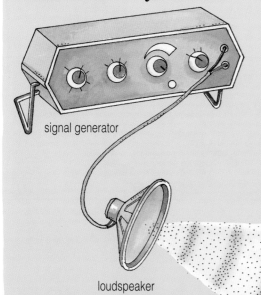

signal generator

loudspeaker

Your teacher will connect a **signal generator** to a **loudspeaker**. The signal generator makes the loudspeaker vibrate, so that it makes a sound wave.

The dial on the signal generator tells you the **frequency** of the vibration. The frequency is measured in **hertz** (also written as **Hz**).

If the frequency is 100 Hz, the loudspeaker is vibrating 100 times in every second. This gives you a low note, like a bass guitar. We say it has a low **pitch**.

- Turn the knob on the signal generator to a higher frequency.
- **j** What happens to the pitch of the note?
- **k** What is the highest frequency you can hear?
- **l** What is the lowest frequency you can hear?

- *Predict* what you will find if you test the hearing range of some adults. Then try it. What do you find?

- Now connect a **microphone** to a **c**athode **r**ay **o**scilloscope (**CRO**).

  A CRO is a special TV set. It shows you the **wave-form** of the sound wave.

- Look at the wave-form as you change the frequency (pitch) of the sound.

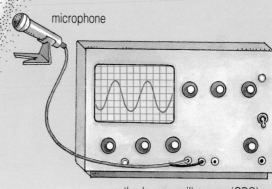

microphone

cathode ray oscilloscope (CRO)

▶ The diagram shows the hearing range of some animals:

m What do you notice?

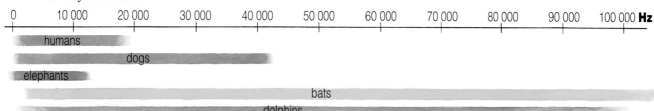

| 0 | 10 000 | 20 000 | 30 000 | 40 000 | 50 000 | 60 000 | 70 000 | 80 000 | 90 000 | 100 000 **Hz** |

humans

dogs

elephants

bats

dolphins

Things to do

1 Copy and complete:
a) Sound can travel through solids, ... , and gases.
b) Sound cannot travel through a . . . .
c) In a sound wave, the tiny . . . . are pushed together and pulled farther apart.
d) The frequency of a vibration is measured in . . . . This is also written . . . .
e) A note with a high pitch has a high . . . .
f) The human range of hearing is from about 20 Hz up to about . . . . Hz.

2 Why can't sound travel:
a) in a vacuum? b) on the Moon?

3 Sound travels at different speeds in different materials:

Material	air	water	wood	iron
Speed of sound (m/s)	330	1500	4000	5000

a) Plot a bar-chart of this information.
b) What pattern can you see?

Musical sounds

▶ Touch the front of your throat while you say 'aaah'.
Can you feel it **vibrating?**

a Explain, step by step, how someone else can hear this sound.

b What is vibrating in 1) a guitar?
2) a drum?
3) the recorder in the photo?

Waves on a spring

Use a 'slinky' spring again to 'model' a sound wave.
As the energy travels along, look carefully at the moving pattern:

← wavelength →

→ energy   This kind of wave is called a **longitudinal** wave.

When you speak, the sound energy travels through the air. It is the **molecules** of air that vibrate:

The energy is transferred from molecule to molecule. They vibrate like the coils of the spring.

c Why can't sound travel through a vacuum?

An oscilloscope

Use a **c**athode **r**ay **o**scilloscope (**CRO**) again to
investigate sound waves.

The sound energy enters the microphone. The energy is
transferred to electrical energy, which goes to the oscilloscope.
A graph of the wave is shown on the screen:

What happens if you 1) turn the Y-shift knob?
2) turn the X-shift knob?
3) switch the time-base off and on?
4) change the Y-gain dial?

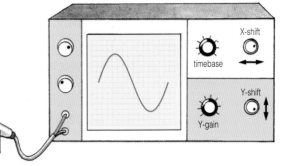

Loudness and amplitude

● Hum or whistle a quiet sound into the microphone.

● Then do the same sound but **louder**.
How does the wave change?

● Sketch the 2 waves.

● How does the **amplitude** of a wave depend on
the loudness of the sound?

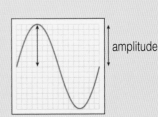

amplitude

Which note is louder?

Pitch and frequency

- Hum or whistle a note with a low pitch, and then a high pitch. How does the wave change?

- Sketch the 2 waves.

wavelength

Low frequency

- Waves with a shorter wavelength have a higher *frequency*. The molecules are vibrating more often. Frequency is measured in **hertz** (**Hz**). A note of 300 Hz means it is vibrating 300 times in each second.

- How does the pitch of your notes depend on the frequency?

- Blow a dog-whistle. What do you notice? What is ultra-sound?

High frequency

Which note is high-pitched?

- Connect a signal-generator to a loudspeaker, to make sound waves of different frequencies. What is the highest note you can hear? Sketch its wave.

Musical instruments

- Make different sounds – aaah, ooo, eee – while you watch the screen.

- Play different musical instruments. Play the same note on each one, and sketch the waves.

- In what ways are the waves   1) the same?   2) different?

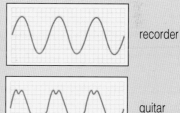

recorder

guitar

violin

1 Copy and complete:
a) A sound . . . . is caused by vibrations. It is a . . . . wave. The energy is transferred from molecule to . . . .
b) A loud sound has a large . . . .
c) A high-pitched sound has a high . . . .
d) Frequency is measured in . . . . (Hz).

2 Write down the names of 10 musical instruments that do not use electricity. For each one, say whether it is plucked, blown, bowed, hit or shaken.

3 Humming birds make a noise by beating their wings very quickly. Plan an investigation to find out the frequency at which their wings vibrate.

4 The diagrams below show 4 waves.
a) Which has the largest amplitude?
b) Which has the highest frequency?
c) Which was the quietest sound?
d) Which sound had the lowest pitch?
e) Which 2 have the same amplitude?
f) Which 2 have the same frequency?

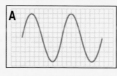

A

B

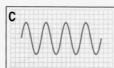

C

D

Things to do

Noise annoys

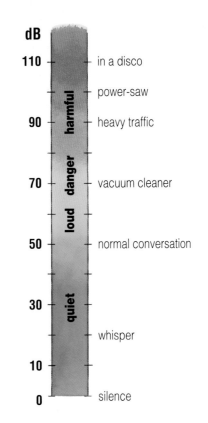

▶ What do you think is the difference between **music** and **noise?**

a Write down 3 words to describe music.
b Write down 3 words to describe noise.

Noise is any sound that we don't like. It is a kind of pollution.

c Give 3 examples of sounds that are noise to you.

The loudness of any sound is measured in **decibels**. This is often written as **dB**.

The quietest sound anyone could hear is zero decibels (0 dB). A louder sound, with more decibels, has more **energy**.

The scale shows the loudness of different sounds:

d What is the loudness, in dB, of a normal conversation?

Try to estimate the loudness (in dB) of these sounds,
e birds singing,
f a food-mixer.

Loud sounds are dangerous. They can make you permanently deaf.

g If you were using a power-saw, what should you wear?

h Why is it often dangerous in a disco?

i Why does the noise in a sports hall sound loud?
j *Why* would it change if there was a carpet and curtains?

dB	
110	in a disco
	power-saw
90	heavy traffic
70	vacuum cleaner
50	normal conversation
30	
	whisper
10	
0	silence

harmful — danger — loud — quiet

Investigating sound-levels

Use a sound-level meter to measure the loudness of some sounds:

Show your results on a scale like the one above.

Using a sound-level meter   Wearing ear-defenders at work

Noisy neighbours

Kelly's neighbours are very noisy. The noise comes through the walls while Kelly is doing her homework. She wants to make the walls of her bedroom sound-proof.

Plan an investigation to find out *which material is best for making sounds quieter*.

- Your teacher will give you several materials.
  For example: paper, kitchen foil, foam rubber, cloth, polythene bag, plasticine, sellotape, cotton wool, etc.

- *Predict* which material you think will be best.
  Can you explain why?

- You will need something to make the sound.
  For example: a clock or a buzzer, or a tin-can containing stones, or a radio, or your voice.

- You will need something to detect how much sound gets through the material.
  For example: your ear, or a sound-level meter, or a microphone and CRO.

materials

sources

detectors

- Plan what you are going to do. How will you make it a *fair and safe test*? How many times will you test each one?
- How will you record your results?
- Show your plan to your teacher, and then do it.

- Write a report for Kelly, explaining which material is best for sound-proofing her room.

- Which sorts of materials are best? Is there any pattern in your results?

- How could you improve your test?

Imagine you live on a very busy road, with a lot of traffic noise.
In your group, discuss these questions:

k How could you cut down the traffic noise in your garden?

l How could you cut down traffic noise inside your house?

m What do you think should be done to reduce traffic noise in towns?

---

1 Copy and complete:
a) The loudness of a sound is measured in . . . . (often written as . . . .).
b) A louder sound has more . . . .

2 Design a poster to encourage teenagers to protect their ears.

3 Imagine you are working in a noisy office. What suggestions could you make to improve it?

4 You are looking out of the window, in a dark room, drinking a cup of coffee, and with the TV on. A deaf person comes into the room. What things would you do before speaking to her?

5 Dave says, "I can work better on my homework if I have some music playing." Wayne says, "I don't agree – the music will spoil your concentration."
Plan an investigation to see who is right.

Things to do

Sound travels

Learn about:
- noise pollution
- the speed of sound

▶ Why do some people wear a hearing aid?
Do you know anyone who wears one?
What do you think is inside a hearing aid?

Your ear can easily be damaged. This causes deafness.
You looked at this earlier.

a In what ways can your ear be damaged?
Write down as much as you can remember about this.

How loud?

If you stand too close to a loudspeaker in a disco, you could damage
your hearing.
Does it matter **where** you stand? Plan an investigation to find out.

- What will you use to detect the sound?
- How will you keep a record of the different positions that you try?
- Predict what you think you will find. If you have time, try it.
- What do you feel after listening to a loud sound for a while?

Noise pollution

The chart shows some data about loudness levels,
which are measured in decibels (dB):

(The loudness levels are a rough guide, but
actual values depend on the exact situation.)

b Continuous noise levels of 90 dB or more
can damage your hearing.
Make a list of any of these situations in your life.

c The loudness shown for a rock concert depends
on whether it is inside a hall or in the open-air.
Why do you think this is?

d The damage depends on the loudness level **and**
the length of time you listen to it.
For how long is it safe to listen (without a break)
to a personal stereo at 110 dB?

e Look at the list of exposure times.
There is a pattern. Describe the pattern you see.

f Professional rock musicians wear expensive ear-
plugs. Why do you think they are each specially
made to be close-fitting in the musician's ear?

Approx. loudness in decibels		Maximum safe exposure time
140	boom stereo in car, jet engine at 100 ft	—
130	rock concert indoors	—
120	rock concert outdoors, loud stereo in car	$7\frac{1}{2}$ mins
110	personal stereo full on; some cinemas	30 mins
100	personal stereo on 6/10 setting	2 hours
90	loud party, motorbike, train	8 hours
80	school canteen, traffic noise in car	
70		
60	conversation	
50		
40		
30	whispering	
20		
10		

Speed of sound in air

▶ Joanne sees a lightning flash on a hill which she knows is 1000 metres away. She hears the sound 3 seconds later.

g Which travels faster, sound or light?

h Use the formula:

$$\text{speed} = \frac{\text{distance travelled}}{\text{time taken}}$$

to calculate the speed of sound in air.

▶ Sound can travel through solids, liquids and gases.

i What can you say about the particles in a solid compared with the particles in a gas?

j Can you use this idea to explain why sound travels faster in a solid than in a gas?

k Do you think sound travels faster in water or in air?

Speed of sound in water

Sailors can use **echoes** to find the depth of the sea:

▶ Suppose this ship sent out a sound wave, and got back an echo after 1 second.
The captain's chart says that the sea is 750 m deep.

l Use the formula:

$$\text{speed} = \frac{\text{distance travelled}}{\text{time taken}}$$

to calculate the speed of sound in water.

m If the fish are 250 m deep, what would be their echo time?

n The boat then moves into very deep water.
Explain why it is harder to detect the echo.

Reflecting a sound

Plan an experiment to see if the angle of incidence is equal to the angle of reflection for a sound wave.

Things to do

1 Copy and complete:
a) Light travels . . . . than sound.
b) The formula for speed is:
c) Sound travels . . . . in iron than in water.
It travels . . . . in water than in air.
d) When sound is reflected, you get an . . . .
This is used in . . . . -sounding.
e) Ear damage depends on the . . . . level and the length of . . . . you listen to it.

2 Make a list of jobs you could not do if your hearing was damaged.

3 If you hear thunder 15 seconds after seeing lightning, how far away is the storm? (Speed of sound = 330 m/s)

4 Write a full page colour advert for a teenage magazine warning of the dangers of excessively loud noise.

5 School canteens are usually very noisy. How could you make yours quieter?
Draw a plan of it and label all the improvements you would make if you were an architect.

Physics at Work

Astronauts

These astronauts can see each other, but they can't hear each other except on the radio.

What does this tell you about:

a light waves?

b sound waves?

c radio waves?

d How would this change if they stood together with their helmets touching?

Whales can 'sing' messages to each other. They can hear each other a hundred miles away through the sea (but not through the air).

e Try to use particle theory (molecules) to explain why sound travels farther in water than in air.

Bats and **dolphins** use ultrasound to find food and 'see' in the dark.

f What is ultrasound?

g How do the bats and dolphins use it?

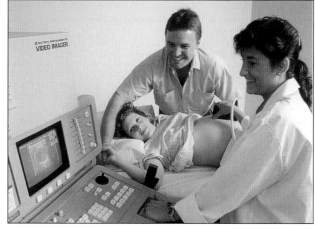

An ultrasonic **scanner** is used to look at a baby inside the mother's womb. It works like the echo-sounder on a ship.

h Explain how you think it works.

Geologists use **echo-sounding** to search for oil and gas.

i In the diagram, which microphone will receive the sound first?

j The speed of sound in rock is about 4000 m/s. If the sound arrives at the microphone after $\frac{1}{2}$ second, estimate the depth of the hard rock layer.

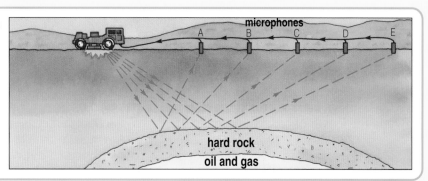

microphones
A   B   C   D   E
hard rock
oil and gas

Ideas about sound

Learn about:
● ideas about sound
● measuring its speed

One of the first experiments on sound was done by **Otto von Guericke** in 1654. He used an air pump to pump air out of a bottle. There was a clockwork bell inside the bottle, and its sound got weaker and weaker as the air was removed.
This shows that sound cannot travel through a vacuum. It needs a medium (a solid, liquid or gas) to pass on the vibrations.

Isaac Newton worked out a formula for the speed of sound in 1698.
The first person to measure the speed of sound accurately was **William Derham** in 1708. He watched and listened as a cannon was fired 19 km (12 miles) away.

Otto von Guericke

Newton's work predicted that sound should travel faster in water than in air. This was proved by an experiment on Lake Geneva in 1827.
An under-water bell was rung at the same time as some gunpowder was lit. On another boat 14 km (9 miles) away, the flash was seen (at night) and the sound heard through the water by a large ear trumpet dipping into the water.

Sound travels about 4 times faster through water than through air. This is because the particles (molecules) of water are held closer together. They spring back together more quickly and pass on the sound pulses more quickly.

---

1 Think about the experiment done by William Derham.
a) Explain in detail what you think he did. What instruments would he need?
b) With a distance of 19 km he found that the sound took 56 seconds. What was the speed of sound in air?

2 Sound travels faster through a solid than through air. Can you explain this?

3 Think about the experiment done at night on Lake Geneva.
a) Sketch what you think the apparatus looked like on each boat.
b) With a distance of 14 km the sound took 10 s. What was the speed of sound in water?

4 Choose one of the scientists and find out more details of his life.

Things to do

Questions

1 Carry out a sound survey among your family and friends.
Make a list of everyone's favourite and least favourite sounds.

2 a) Copy out the table. Tick a column to show if the
frequency is a high or a low pitch.
b) Older people cannot hear very high notes. Why do you
think your range of hearing will get smaller as you get older?
c) What is ultra-sound? How do bats use it?

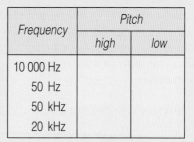

Frequency	Pitch	
	high	low
10 000 Hz		
50 Hz		
50 kHz		
20 kHz		

3 a) Imagine, in a thunderstorm, lightning strikes 660 metres away
from you. Describe and explain what you would observe.
(The speed of sound in air is 330 metres per second.)
b) A fighter plane flies at a speed of Mach 4. The speed of sound
is called Mach 1. How fast is the plane flying?

4 Plan an investigation to see if children have better hearing than
adults. How would you make it a fair test?

5 Alan, Bev and Claire have read that people find
high frequency sounds more annoying than low
frequency sounds. They each have a hypothesis:

Alan says, "I think it's because high frequency sounds
make it harder to hear someone talking."

Bev says, "I think it's because our ears are more
sensitive to high-pitched sounds."

Claire says, "I think it's because high frequency sounds
can penetrate through walls more easily."

a) Do you agree with any of these hypotheses?
b) Choose one, and plan an investigation to test it.

6 Kelly says, "There ought to be a law against playing music
loudly." Do you agree with her?
Give your arguments for and against this idea.

7 The diagram shows a wave-form on an oscilloscope:
a) What is the time taken for 1 wave?
b) How many waves are there in 100 milliseconds?
c) How many waves will there be in 1 second?
(1 second = 1000 milliseconds)
d) What is the frequency of the wave?
e) Would this be a high note or a low note?

8 The time-keeper of a 100 m race stands at the finishing line.
He starts his stop-watch when he hears the starter's pistol.

a) Will the time he measures be too long or too short?
b) By how much? (Speed of sound in air = 340 m/s)

Glossary

Absorb
When light, sound or another form of energy is taken in by something, e.g. black paper absorbs light energy, or when digested food is absorbed into the blood from the small intestine.

Adaptation
A feature that helps a plant or animal to survive in changing conditions.

Amplitude
The size of a vibration or wave, measured from its mid-point. A loud sound has a large amplitude.

Antibiotic
A useful drug that helps your body fight a disease.

Artery
A blood vessel that carries blood away from the heart.

Atom
The smallest part of an element.

Bacteria
Microbes made up of one cell, visible with a microscope. Bacteria can grow quickly and some of them cause disease, e.g. pneumonia.

Boiling point
The temperature at which a liquid boils and changes into a gas.

Bronchus
One of the tubes at the bottom of the wind-pipe (trachea) that lead to the lungs.

Capillaries
Tiny blood vessels that let substances like oxygen, food and carbon dioxide pass into and out of the blood.

Carbohydrate
Your body's fuel. Food like glucose that gives you your energy.

Competition
A struggle for survival. Living things compete for scarce resources, e.g. space.

Compound
A substance made when 2 or more elements are chemically joined together, e.g. water is a compound made from hydrogen and oxygen.

Conductor
An electrical conductor allows a current to flow through it. A thermal conductor allows heat energy to pass through it. All metals are good conductors.

Convection
The transfer of heat by currents in a liquid or a gas.

Digestion
Breaking down food so that it is small enough to pass through the gut into the blood.

Dispersion
The splitting of a beam of white light into the 7 colours of the spectrum, by passing it through a prism.

Distillation
A way to separate a liquid from a mixture of liquids, by boiling off the substances at different temperatures.

Electro-magnet
A coil of wire becomes a magnet when a current flows through it.

Element
A substance that is made of only one type of atom.

Enzymes
Chemicals that act like catalysts to speed up digestion of our food.

Equation
A shorthand way of showing the changes that take place in a chemical reaction
e.g.   iron + sulphur $\rightarrow$ iron sulphide
        Fe   +   S $\rightarrow$ FeS

Erosion
The wearing away of rocks.

Fat
Food used as a store of energy and to insulate our bodies so we lose less heat.

Formula
A combination of symbols to show the elements which a compound contains,
e.g. MgO is the formula for magnesium oxide.

Fossil
The remains of an animal or plant which have been preserved in rocks.

Frequency
The number of complete vibrations in each second. A sound with a high frequency has a high pitch.

Fungi
Moulds, such as yeast or mushrooms, that produce spores.

Group
All the elements in one column down the periodic table.

Igneous rock
A rock formed by molten (melted) material cooling.

Image
When you look in a mirror, you see an image of yourself.

Immune
Not being able to catch a particular disease because you have the antibodies in your blood to fight it.

Insulator
An electrical insulator does not allow a current to flow easily. A thermal insulator does not let heat energy flow easily.

Intestine
Tube below the stomach where food is digested and absorbed.

Law of reflection
When light rays bounce off a mirror:
angle of incidence = angle of reflection.

Lava
Molten rock ejected from a volcano.

Lungs
The organs in our body that collect oxygen and get rid of carbon dioxide.

Magma
Hot molten rock below the Earth's surface.

Magnetic field
The area round a magnet where it attracts or repels another magnet.

Melting point
The temperature at which a solid melts and changes into a liquid.

Metamorphic rock
A rock formed by heating and compressing (squeezing) an existing rock.

Mixture
A substance made when some elements or compounds are mixed together. It is *not* a pure substance.

Molecule
A group of atoms joined together.

Non-metal
An element which does not conduct electricity. (The exception to this is graphite – a form of carbon which is a non-metal, but it does conduct).

Opaque
An opaque object will not let light pass through it.

Periodic table
An arrangement of elements in the order of their atomic numbers, forming groups and periods.

Pitch
A whistle has a high pitch, a bass guitar has a low pitch.

Population
A group of animals or plants of the same species living in the same habitat.

Porosity
The ability to absorb a liquid such as water, e.g. sandstone is a porous rock.

Protein
Food needed for growth and repair of cells.

Pyramid of numbers
A diagram to show how many living things there are at each level in a food chain.

Radiation
Rays of light, X-rays, radio waves, etc., including the transfer of energy through a vacuum.

Reflection
When light bounces off an object.

Refraction
A ray of light passing from one substance into another is bent (refracted).

Relay
A switch that is operated by an electro-magnet. A small current can switch on a large current.

Respiration
The release of energy from food in our cells. Usually using up oxygen and producing carbon dioxide.

glucose + oxygen ➡ carbon dioxide + water + energy

Rock cycle
A cycle that means that one type of rock can be changed into another type of rock over a period of time.

Scattering
When rays of light hit a rough surface (like paper) they reflect off in all directions.

Sedimentary rock
A rock formed by squashing together layers of material (sediments) that settle out in water.

Thermal transfer
Heat can be transferred by conduction, convection, radiation and evaporation.

Trachea
The wind-pipe taking air to and from the lungs.

Vein
A blood vessel that carries blood back to the heart.

Vibrating
Moving backwards and forwards quickly, e.g. the particles in a solid vibrate.

Viruses
Extremely small microbes which are not visible with a microscope. Many viruses spread disease by invading cells and copying themselves, e.g. influenza.

Wavelength
The distance between 2 peaks of a wave.

Weathering
The crumbling away of rocks caused by weather conditions such as rain and extreme changes in temperature.

Index